T0306094

# Small-scale Computational Vibration of Carbon Nanotubes: Composite Structure

# RIVER PUBLISHERS SERIES IN MATHEMATICAL, STATISTICAL AND COMPUTATIONAL MODELLING FOR ENGINEERING

*Series Editors:*

**MANGEY RAM**
*Graphic Era University, India*

**TADASHI DOHI**
*Hiroshima University, Japan*

**ALIAKBAR MONTAZER HAGHIGHI**
*Prairie View Texas A& M University, USA*

Applied mathematical techniques along with statistical and computational data analysis has become vital skills across the physical sciences. The purpose of this book series is to present novel applications of numerical and computational modelling and data analysis across the applied sciences. We encourage applied mathematicians, statisticians, data scientists and computing engineers working in a comprehensive range of research fields to showcase different techniques and skills, such as differential equations, finite element method, algorithms, discrete mathematics, numerical simulation, machine learning, probability and statistics, fuzzy theory, etc.

Books published in the series include professional research monographs, edited volumes, conference proceedings, handbooks and textbooks, which provide new insights for researchers, specialists in industry, and graduate students.

Topics included in this series are as follows:-

- Discrete mathematics and computation
- Fault diagnosis and fault tolerance
- Finite element method (FEM) modeling/simulation
- Fuzzy and possibility theory
- Fuzzy logic and neuro-fuzzy systems for relevant engineering applications
- Game Theory
- Mathematical concepts and applications
- Modelling in engineering applications
- Numerical simulations
- Optimization and algorithms
- Queueing systems
- Resilience
- Stochastic modelling and statistical inference
- Stochastic Processes
- Structural Mechanics
- Theoretical and applied mechanics

For a list of other books in this series, visit www.riverpublishers.com

# Small-scale Computational Vibration of Carbon Nanotubes: Composite Structure

**Muzamal Hussain**

University of Sahiwal, Sahiwal, Pakistan
Government College University Faisalabad, Pakistan

Routledge
Taylor & Francis Group
NEW YORK AND LONDON

**Published 2024 by River Publishers**
River Publishers
Alsbjergvej 10, 9260 Gistrup, Denmark
www.riverpublishers.com

**Distributed exclusively by Routledge**
605 Third Avenue, New York, NY 10017, USA
4 Park Square, Milton Park, Abingdon, Oxon OX14 4RN

*Small-scale Computational Vibration of Carbon Nanotubes: Composite Structure / by* Muzamal Hussain.

Routledge is an imprint of the Taylor & Francis Group, an informa business

ISBN 978-87-7022-865-7 (hardback)
ISBN 978-87-7004-065-5 (paperback)
ISBN 978-10-0382-340-7 (online)
ISBN 978-10-3265-622-9 (master ebook)

While every effort is made to provide dependable information, the publisher, authors, and editors cannot be held responsible for any errors or omissions.

## Dedication

*I would like to dedicate this Book to the memory of my father (Late), to the inspiration of my mother, to the love and support of my family.*

# Contents

# Preface

Due to exciting and challenging research, I am tremendously fortunate to be convoluted in writing this book. It has enriched my life, giving me an opportunity to work. As I started the book from scratch, my thinking and understanding capability is increased on accomplishing. I consider myself extremely lucky to be able to work under guidance of such a dynamic national and international researchers. The author wrote this book on the vibration of carbon nanotubes. Actually, carbon nanotubes have been firstly reported in 1991 by Sumio Iijima. The discovery of CNTs opens a new window to the researchers in different fields. The vibration analysis of the nano-structures is a new and very hot topic, and accordingly, there are many unsolved and vague problems for the future researchers in this area. The nanotubes are used for a number of scientific and technological applications. With a vast area of potential innovation, however CNTs demands more understanding to investigate its mechanical properties.

The vibration characteristic of SWCNTs using continuum analyses is investigated to explore compression, torsion, bending and axial tension. It is found that with small change in length, height, Poisson's ratio, density and stiffness, the CNTs are very sensitive and with smaller radii, the tubes have higher sensitivities and the natural frequency upshifts and downshifts respectively. Various boundary conditions are utilized and have momentous effect on the deformation of the tubes.

To assess the frequency confirmation carried out in this study are compared with the earlier computations. The computer software MATLAB is used for the extraction of results. Convergence of present study is done to view the accuracy. With properly chosen parameters, the present frequencies are compared with beam element, Timoshenko beam model, molecular dynamic (MD) simulations, DTM and Bubnov–Galerkin method, Resonant Raman Scattering (RRS) and Raman Spectroscopy.

I have completed this book according to the basic knowledge and apply mathematical techniques for growing researchers in the field of vibration of tubes/shells/plates. Many geometrical and physical parameters are demonstrated for basic explanation and understandings.

The theoretical results obtained from this book can be used for the sensitivity analysis of the nano- and bio-sensors as a practical material. The proposed models of the present work can be used for the vibrations analysis with experimental set.

**Muzamal Hussain**
University of Sahiwal, Sahiwal, Pakistan
Government College University Faisalabad, Pakistan

# Acknowledgments

I hereby declare that the title of book "**Small-scale Computational Vibration of Carbon Nanotubes: Composite Structure**" and the contents of the book are the product of my own research, and no part has been copied from any published source (except the reference, standard mathematical equations/formulas/protocols, etc.). I further declare that this work has not been submitted for the award of any other degree/diploma.

**Muzamal Hussain**

# List of Figures

# List of Tables

# List of Notations

| | |
|---|---|
| $C_k$ | Chiral vector |
| $u(x, \theta, t),$ | |
| $v(x, \theta, t),$ | Displacement deformations |
| $w(x, \theta, t),$ | |
| $\rho$ | Mass density |
| $E$ | Young's modulus |
| $v$ | Poisson's ratio |
| $h$ | Wall thickness |
| $L$ | length |
| $d$ | diameter |
| $(x, \theta, z)$ | Axial, circumferential, and radial coordinates |
| $\sigma$ and $e$ | Stress and strain vectors |
| $Q$ | Reduced stiffness matrix |
| $(e_1, e_2, \gamma)$ | References surface strains |
| $(k_1, k_2, \tau)$ | References surface curvatures |
| $N$ and $M$ | Force and moment components |
| $A, B$ and $D$ | The extensional, coupling and bending stiffness Matrices |
| $U$ | Strain energy |
| $T$ | kinetic energy |
| $\Pi$ | Lagrangian expression |
| $A_m, B_m, C_m$ | The amplitudes of the vibrations |
| $\omega$ | natural frequency |
| $L/d$ | Ratio of length to diameter |
| $f(\text{THz})$ | Frequency in Tera Hertz |
| $(\alpha, \beta, \gamma)$ | The dimensionless coordinates |
| $(\varepsilon_\alpha, \varepsilon_\beta, \varepsilon_{\alpha\beta})$ | The dimensionless strains |
| $(\sigma_\alpha, \sigma_\beta, \sigma_{\alpha\beta})$ | The dimensionless stresses |
| $(\nabla^2)$ | Laplace operator |
| $h_0$ | The effective thickness |
| $h$ | The equivalent thickness |

| | |
|---|---|
| $K$ | Elastic coefficient |
| $\eta$ | Viscous coefficient |
| $D$ | Effective bending stiffness |
| $Eh$ | In-plane rigidity |
| $\rho h$ | Mass density per unit lateral area |
| $R$ | Radius |
| $h/d$ | Ratio of height-to-diameter |
| $\gamma(x)$ | mode shape |
| $I$ | moment of inertia |
| $A$ | The cross-section area |
| $G$ | Shear modulus |

# List of Abbreviations

| | |
|---|---|
| 2-D | 2-dimensional |
| 3-D | 3-dimensional |
| BCs | Boundary conditions |
| BGM | Bubnov–Galerkin method |
| C-60 | Carbon 60 |
| C-C | Clamped-clamped |
| C-F | Clamped-free |
| CNTs | Carbon nanotubes |
| CSs | Cylindrical shell |
| C-SS | Clamped-simply supported |
| DQM | Differential quadrature method |
| DSC | Discrete singular convolution |
| DST | Donnell's shell theory |
| DTM | Differential transform method |
| DWCNTs | Double-walled carbon nanotubes |
| EBM | Euler beam model |
| EBT | Eringen's beam theory |
| FST | Flügge shell theory |
| GLT | Galerkin's technique |
| GPa.nm | Giga Pascal nanometer |
| GT | Galerkin's technique |
| HPM | Homotopy perturbation method |
| LGF | Lagrangian functional |
| MATLAB | Matrix Laboratory |
| MD | Molecular dynamic |
| MWCNTs | Multi-walled carbon nanotubes |
| ODEs | Ordinary differential equations |
| PDEs | Partial differential equations |
| RKM | Rung–Kutta method |
| RRM | Rayleigh–Ritz method |
| RRS | Resonant Raman scattering |

| | |
|---|---|
| SGT | Strain gradient theory |
| SS-SS | Simply supported-simply supported |
| SWCNTs | Single-walled carbon nanotubes |
| TBM | Timoshenko beam model |
| THz | Terahertz |
| WPA | Wave propagation approach |
| WTM | Winkler-type model |

# Abstract

For the last decade, free vibration analyses of CNTs have been an influential aspect of dynamic science. Carbon nanotubes (CNTs) were discovered by Iijima (1991), which may be used in a variety of fields like material reinforcement, aerospace, medicine, defense, and microelectronic devices. Extensive research has been conducted in the field of nanotubes, and as its use in commercial applications has increased, rapid development has occurred. Nanotubes are used for a number of scientific and technological applications. Owing to the small size of the micro beam, these structures are very appropriate for designing small instruments. Large deformation of CNTs related to morphological design corresponds to the release of energy in strains–stress curve.

The present work investigates vibrations of single-walled carbon nanotubes (SWCNTs). There are three categories of single-walled carbon nanotubes viz armchair, zigzag, and chiral are used here for their vibration characteristics with different parameters and boundary conditions. The vibrations are established using the Rayleigh–Ritz method, orthotropic shell model, Donnell's shell theory, Sander's shell theory, and Euler theory formulation. These proposed models are quite straightforward for the vibrational analysis of these structures of SWCNTs.

The vibrations of single-walled carbon nanotubes (SWCNTs) are investigated based on Flügge shell theory. The Rayleigh–Ritz method determines eigenfrequencies for single-walled carbon nanotubes. The solution is obtained using the geometric characteristics and boundary conditions for the natural frequencies of SWCNTs. Influences of length-to-diameter ratios and the two boundary conditions on the natural frequencies of armchair, zigzag, and chiral SWCNTs are examined. The natural frequencies decrease as the length-to-diameter ratio increases, and the effect of frequencies is less significant and more prominent for a long tube.

Furthermore, the vibration characteristics of single-walled carbon nanotubes (SWCNTs) are developed using an orthotropic shell model. Based on this model, the frequency effect of stiffness with boundary conditions is discussed and examined. The frequencies are calculated against the stiffness with two prescribed boundary conditions. The frequencies increase on increasing the stiffness of the carbon nanotubes. Then, the frequency jumps vertically upward. These sudden jumps show that the material of the tube is very stiff during vibration. It can be seen that the frequencies of clamped-free boundary conditions are lower than simply supported condition.

Moreover, vibration characteristics of SWCNTs based on Donnell's shell theory are developed. The governing equation of motion and boundary conditions using the wave propagation approach is written in the form of eigenvalue to extract the frequencies of CNTs. The effect of height-to-diameter ratios and boundary conditions for SWCNTs on the frequencies are calculated. Based on Donnell's shell theory, it is observed that with the increase in the height-to-diameter ratios, the frequencies increase for all tubes. The frequency pattern with two prescribed boundary conditions seems parallel for overall height-to-diameter ratio values. It is also seen that the frequency curves with changing the values of the height-to-diameter ratio of C-F boundary condition are the lowest outcome.

An analysis of SWCNTs based on Sander's shell theory is carried out. The effect of Poisson's ratio, boundary conditions, and different chirality of carbon nanotubes SWCNTs is discussed. A change of natural frequency for different Poisson's ratio of SWCNTs is conducted. Moreover, this model predicts the phenomena of frequencies of chirality of SWCNTs. With the decrease in ratios of Poisson, the frequency increases. Poisson's ratio directly measures the deformation in the material. The frequency value increases with the increase of single-walled carbon nanotubes indices. The prescribed boundary conditions used are simply supported and clamped-simply supported.

Based on Galerkin's method, a comprehensive estimation of modified Euler-Bernoulli beam model for vibration of single-walled carbon nanotubes (DWCNTs). The two different immovable boundary conditions are applied at the end of these tubes. The impact of density on natural frequency is investigated. The vibration frequencies are inversely proportional to the mass of the zigzag single-walled carbon nanotubes. So due to increase in density results a decrease in resonant frequency.

To assess the frequency, confirmation carried out in this study is compared with the earlier computations. The computer software MATLAB is used for the extraction of results. Convergence of the present study is done to view the accuracy. With properly chosen parameters, the present frequencies are compared with beam element, Timoshenko beam model (TBM), molecular dynamic (MD) simulations, differential transform method (DTM) and Bubnov–Galerkin method, Resonant Raman Scattering (RRS) and Raman Spectroscopy. The present results are validated with the exact and experimental results in tabular form. The comparisons show that the vibration responses of SWCNTs are influenced by frequency variation with length-to-diameter ratios, stiffness, height-to-diameter ratios, and Poisson's ratio.

Moreover, the index order of the armchair, zigzag, and chiral tube impresses the frequency values. So the effect of length-to-diameter ratios is seen for armchair (4, 4), (11, 11), zigzag (6, 0), (13, 0), and chiral (7, 4), (12, 6), the influence of stiffness is considered for (5, 5), (7, 7), and (12, 12), zigzag (6, 0), (9, 0), (14, 0), and chiral (6, 2), (9, 4), (12, 7), the impact of height-to-diameter ratios, boundary conditions for armchair (4, 4), (11, 11), zigzag (6, 0), (13, 0), and chiral (7, 4), (12, 6), the effect of Poisson's ratio, armchair (6, 6), (8, 8), (10, 10), zigzag (7, 0), (11, 0), (15, 0), and chiral (5, 3), (8, 2), (11, 7), the influence of armchair (3, 3), (9, 9), (13, 13), zigzag (4, 0), (8, 0), (10, 0), and chiral (4, 2), (7, 3), (10, 5) single-walled carbon nanotubes on the frequencies.

The boundary conditions are used in this study as follows for length-to-diameter ratios (clamped-clamped and simply supported-simply supported), for stiffness (simply supported-simply supported and clamped-free), for height-to-radius ratios (clamped-clamped and clamped-free), for Poisson's ratio (simply supported-simply supported and clamped-simply supported), for density (clamped-clamped and clamped-free). The symmetric frequency for BCs is seen as C-C > SS-SS > C-SS > C-F for the overall study. These boundary conditions have a momentous effect on the deformation of the tubes. The frequencies are presented in THz throughout the paper.

With a vast area of potential innovation, however, CNTs demand more understanding to investigate their mechanical properties.

# 1

# Introduction

## 1.1 Carbon Nanotube

Carbon is one of the most versatile elements that is found in nature, owing to exceptional characteristics, potential applications, and substantial impressions on the industry. Carbon is found in its miscellaneous forms of diamond, graphite, and polymers, which are very important for our lives. Many earliest cave paintings at Altamira and Lascaux were produced with a mixture of charcoal and soot. Carbon black (a better form of soot), graphite, and charcoal have been used for printing, writing, and drawing materials. Carbon black is also largely utilized in the photocopier toner. In the eighteenth century, when charcoal was replaced with coke, this development was very helpful in stimulating the industrial revolution.

Drexler says, "Coal and diamonds, sand and computer chips, cancer and the health organization: throughout history, changes in the arrangement of atoms have differentiated from cheap, diseased and healthy ones. The atoms make up a mixture of soil, air, and waterways: arrange others to make ripe strawberries. They arrange a way to make up for the home and fresh air, and arrange for them to make up for dust and smoke! Our ability to arrange atoms lies in the basics of nanotechnology."

Demand for graphite developed with the electrical industry's growth in the late nineteenth century. In 1896, the first graphite synthetic was produced by the American Edward Acheson. The significance of carbon grew progressively in air supplies and purifying water in the twentieth century. In the 1950s, material researchers invented a new ultra-strong lightweight material known as carbon fiber. However, in the early 1980s, carbon science explored a new horizon, mature disciplines, and major surprises, let alone Nobel Prizes. Kroto et al. [1] and his coworkers discovered a new form of the carbon molecule, Buckminster Fullerene, $C_{60}$, resembling the architectural designs [2]. Owing to this discovery, synthesizing fullerene with carbon nanotubes

1

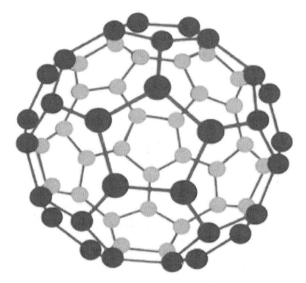

**Figure 1.1**   $C_{60}$: buckminsterfullerene.

suddenly became easy and fashionable. Macromolecules of carbon atoms are arranged at the apexes of pentagons and hexagons, as shown in Figure 1.1.

In 1952, Radushkevich and Lukyanovich [3] observed carbon nanotubes for the first time. At first, however, further progress was very slow. Krätschmer et al. [4] discovered a new form of solid carbon by using infrared spectra and X-ray diffraction with Buckminster Fullerene, $C_{60}$, which excited an inundation of research. Perhaps, CNTs are the most important and wonderful pods of this research. Japanese microscopist Sumio Iijima was fascinated by Krätschmer et al. [4]], working at the NEC laboratories. In a meeting in Richmond, Virginia, Iijima [5] showed beautiful images of CNTs.

The prefix "nano" ($10^{-9}$ m) is the basic unit of length, as small as 100–1000 times smaller than that of biological bacteria or cells. At this scale, the dimensions of the devices and materials begin to range from 10 to 100 atoms. As a result, new trends in physical and chemical effects are observed; and based on the resulting miniaturization or so-called "nanoization" technology, the next generation of cutting-edge products may emerge.

## 1.2 Structure of CNTs

There is a close structural relationship between fullerenes and nanotubes. Considering two "prototype" CNTs, it can be formed by half-cleaving a

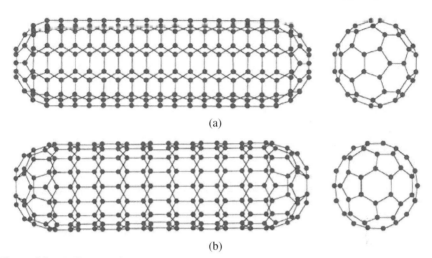

(a)

(b)

**Figure 1.2** A diagram of two nanotubes that can be capped by half of the C-60 molecule [7] (**a**) Armchair (5, 5) (**b**) zigzag (9, 0) structure (see Chapters 3, 4, and 5 for an explanation of the indices).

$C_{60}$ molecule and placing a graphene cylinder between the halves. Armchair nanotubes are created by bisecting $C_{60}$ along one of five axes and forming zigzag nanotubes by separating $C_{60}$ from one of the three axes in parallel. In Figure 1.2, the arrangement of hexagons around the circumference refers to armchairs, and zigzags are shown. The third structure of the nanotube occurs when the hexagons are spirally arranged around the tube axis. More complex cap structures are often observed due to a heptagon and pentagonal carbon ring [6].

## 1.3 Types of CNTs

Basically, nanotubes are two types depending upon the layers, as shown in Figure 1.3. They are

1. Single-walled carbon nanotube
2. Multi-walled carbon nanotube

Some researchers have independently reported the synthesis of SWCNTs [8, 9]. SWCNTs can be prepared by rolling a single graphene layer with hexagonal cells to form the structure of cylindrical fullerene. Since the structure of SWCNTs has appeared approximate to the ideal nanotubes, as

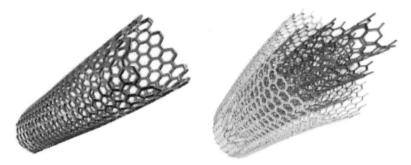

**Figure 1.3**    Single- and multi-walled nanotubes.

shown in Figure 1.2, their research proved an extremely important development in carbon sciences. Furthermore, they proved that these nanotubes have extraordinary properties, and based on their research, many papers on single-walled CNTs have been published. Using their important outcomes, Smalley's group [10] described the synthesis of SWCNTs, the consequences of which are fruitful.

In 1956, Richard Feynman opened a new idea for the researchers in his most famous presentation as "manipulating and controlling things on a small scale". After his most valued talk, the formation of graphitic whiskers, rolling over of graphite sheets, in a form now known as multiwall-carbon nanotubes (MWCNTs) [11–13], as shown in Figure 1.3. They also tend to form bundles or "ropes." Hutchison et al. [14] obtained DWCNTs by arc discharge technique. It was revealed that the inner and outer diameters of DWCNTs are in the range of 1.1–4.2 nm and 1.9–5 nm, respectively, with high-resolution electron microscopy. Another form of CNTs, called MWCNTs, becomes on rolling of more than one layer of graphene. In literature, there are two main models for the structure of MWCNTs: Parchment and Russian. A coaxially single graphite layer is wrapped around itself in the Parchment model, while the Russian model suggests a structure-like scroll for CNTs. The measured distance on closing these layers of graphite is 0.34 nm. Iijima [5] and his coworkers discovered multi-walled carbon nanotubes with several nested cylinders with interlayer spacing.

## 1.4 Properties of Carbon Nanotubes

Further research that possessed amazing physical properties of CNTs has exposed that carbon nanotubes (CNTs) have maximum tensile strength and strain [15–21]. Furthermore, since the amount of hydrogen adsorbed in carbon nanotubes is surprisingly high and reversible, it is possible to use carbon nanotubes as high-capacity hydrogen storage [22].

## 1.5 Categories of CNT

Nanotubes can be categorized into three sorts according to their structure. They are

1. Armchair
2. Zigzag
3. Chiral

Carbon nanotubes (CNTs) fascinate new materials with astonishing mechanical, optical, and electrical properties [23]. They are generated by rolling of the graphene sheet [5, 24]. Carbon nanotube sheets include hexagonal cells that are ideally cut to produce carbon atoms of the tube. In fact, CNTs are kinds of rolled graphene sheets, and the rolling manner shows the tube's basic properties, which is actually the main reason for the extraordinary feature of the CNTs [25]. These cylindrical structures have many fascinating and valuable properties with several potential applications in different fields [24]. The armchair and zigzag carbon nanotubes are usually presented by $(n, n)$ and $(n, 0)$, respectively. The third type of spiral conformation is called chiral [25, 26]. In fact, the index $n$ represents the number of hexagonal lattices associated with the diametric ring of the CNTs, and the index $m$ represents the slope of the hexagonal lattice aligned along the longitudinal axis of the particular nanotube [25]. The translation indices are represented as a pair of integers $(n, m)$, and the diameter of the CNTs can be determined. These integers are labeled on the graphene sheet [23], as shown in Figure 1.4. There are three categories of CNTs, armchairs, zigzags, and chirals, as shown in Figure 1.4 [24].

## 1.6 Chirality Vector of Carbon Nanotubes

Figure 1.1 shows the different rolling style of the graphene sheet. According to this rolling style, three classifications occurred: armchair, zigzag, and chiral. The atomic structure is sketched in Figure 1.2, representing chiral vector $C_K = (n, m)$ where the integers represents the counting of steps along vectors $a_1$ and $a_2$. This chirality of vector determines the carbon nanotubes either is semiconductor or metallic. Moreover, Figure 1.5 shows the orientation of the

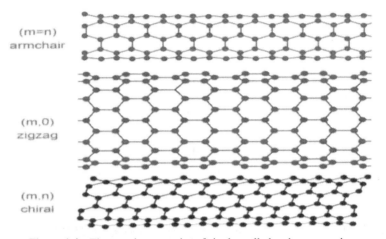

**Figure 1.4**   Three main categories of single-walled carbon nanotube.

graphene sheet as nanotubes become armchair, if $m$ = n, is classified as zigzag if $m$ = 0, and the nanotubes are chiral if $n \neq m$ respectively.

## 1.7 Outline of the Present Study

This book is written based on research principles; in an integrated form, and each chapter is written independently with introductory notes, literature review, mathematical formulation, solution methodology, results and discussion, conclusions, references, and finally last chapter is recommended for future work. This book chapter contains a lot of literature reviews and more than 100 papers as presented by many researchers, and in general, this book chapter provides sound knowledge on basic concepts and huge knowledge for some of engineering mechanics and the scientific community. This vibration investigation is the main basis implementation of SWCNTs in nanotechnology as material reinforcement, aerospace, medicine, defense, and microelectronic devices.

In **Chapter 1**, carbon nanotube, structure of CNTs, types of CNTs, properties of carbon nanotubes, categories of CNT, and chirality vector of carbon nanotubes is presented, starting with extensive understanding report on basic and formal information of carbon nanotubes.

**Chapter 2** aims to study Flügge shell theory based on the Rayleigh–Ritz method. The simplicity of this model is due to its derived attractive results.

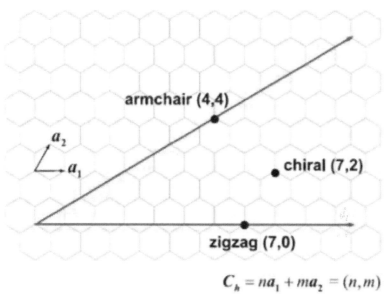

$$C_h = na_1 + ma_2 = (n,m)$$

**Figure 1.5**   Schematic diagram of the chiral vector.

For two sets of boundary conditions clamped-clamped and simply supported-simply supported, are adopted for the frequency behavior of three categories viz armchair, zigzag, and chiral of SWCNTs. According to these two boundary conditions, the nanotubes are clamped on both sides as one end condition, and for the other end condition, the nanotubes are clamped from one end, and the other end is simply supported. Excellent validation is observed with numerical and experimental results. The natural frequencies are affected by length-to-diameter ratios with two boundary conditions.

**Chapter 3** deals with the orthotropic shell model to investigate the vibration response of armchair, zigzag, and chiral SWCNTs. The wave propagation approach (WPA) is developed to generate eigenvalue form with the help of the axial modal function in matrix representation. This study is developed to check the vibration of that system in increasing or decreasing the effect of stiffness with frequencies, which is overcome by applying the boundary conditions. Here only two boundary conditions viz simply supported and clamped-free boundary are used for maintaining the stiff of armchair, zigzag, and chiral SWCNTs. The framework of this model is to check the frequency effect of increasing the stiffness of the material.

The particular motive of **Chapter 4** is to develop the vibration character-istics of armchair, zigzag, and chiral single-walled carbon nanotubes based on Donnell's shell theory (DST). The governing equation of motion and boundary conditions using the wave propagation approach is written in the form of eigenvalue to extract the frequencies of CNTs (armchair, zigzag, and chiral). The chirality of armchair, zigzag, and chiral SWCNTs do not affect the vibration frequencies because the frequencies increase on increasing the height-to-diameter ratios. These variations of frequencies are seen with clamped-clamped and clamped-free end conditions. Eigensolutions of the frequency equation have been determined by writing in MATLAB coding.

**Chapter 5** is an original method of Sander's shell theory to analyze the effects of Poisson's ratios. The two different boundary conditions such as simply supported and clamped-simply supported in order to investigate the natural frequency in detail of these chirality's. The suggested technique to investigate the solution of fundamental eigen relations is Galerkins method, which is an eminent and effectual method to form the fundamental frequency equations. The frequency value of armchair, zigzag, and chiral tubes indicates that the addition of the Poisson effect increases the effective stiffness of single-walled carbon nanotubes.

**Chapter 6** aims to introduce the Euler beam model for vibrations of single-walled CNTs. Also, the frequency characteristics of CNTs for the variation of density versus natural frequency are established. The analysis combines clamped-clamped and clamped-free boundary conditions with different chiral indices. After presenting the convergence of the presented model, the influ-ence of two boundary conditions, density variation, and natural frequency, on armchair, zigzag, and chiral characteristics are investigated in detail.

Finally, **Chapter 7** indicated a general conclusion and main contributions of this book to the mathematicians and material researches. Also, according to these developed models presented in this book to consider for future works.

## References

[1] Kroto, H. W., Heath, J. R., OBrien, S. C., Curl, R. F., & Smalley, R. E. (1985). C60 buckyminsterfullerence. *Nature*, 318, 162.
[2] Fuller, R. B., (1970). Utopia or Oblivion. Bantam books, New York.

[3] Radushkevich, L. V. & Lukyanovich, V. M. (1952). O strukture ugleroda, obrazujucegosja pri termiceskom razlozenii okisi ugleroda na zeleznom kontakte. *Zurn Fisic Chim*, 26(1), 88–95.

[4] Krätschmer, W., Lamb, L. D., Fostiropoulos, K., & Huffman, D. R. (1990). Solid C60: a new form of carbon. *Nature*, 347(6291), 354.

[5] Iijima, S. (1991). Helical microtubules of graphitic carbon. *Nature*, 354 (7), 56–58.

[6] Iijima, S., Ichihashi, T., & Ando, Y. (1992). Pentagons, heptagons and negative curvature in graphitic microtubule growth. *Nature*, 356, 776.

[7] Ge, M., & Sattler, K. (1994). Scanning tunnelling microscopy of single-shell nanotubes of carbon. *Appl. Phys. Lett.*, 65, 2284.

[8] Iijima, S., & Ichihashi, T. (1993). Single-shell carbon nanotubes of 1-nm diameter. *nature*, 363(6430), 603-605.

[9] Bethune, D. S., Kiang, C. H., De Vries, M. S., Gorman, G., Savoy, R., Vazquez, J., & Beyers, R. (1993). Cobalt-catalysed growth of carbon nanotubes with single-atomic-layer walls. *Nature*, 363(6430), 605.

[10] Smalley, R. E., Li, Y., Moore, V. C., Price, B. C., Colorado, Jr, R., Schmidt, H. K., Hauge, R. H., Barron, A. R., & Tour, J. M. (2006). Single Wall Carbon Nanotube Amplification: En Route to a Type-Specific Growth Mechanism. *J Am Chem Soc,* 128, 15824–15829.

[11] Feynman, R. P. (1960). There's plenty of room at the bottom. *Engineering and science*, 2(5), 22–36, 1960.

[12] Bacon, R. (1960). Growth, structure, and properties of graphite whiskers. *Journal of Applied Physics*, 31(2), 283–290, 1960.

[13] Bollmann, W., & Spreadborough, J. (1960). Action of graphite as a lubricant. *Nature*, 186, 29–30.

[14] Hutchison, J. L., Kiselev, N.A., Krinichnaya, E.P., Krestinin, A.V., Loutfy, R.O., Morawsky, A.P., Muradyan, V.E., Obraztsova, E. D., Sloan, J., Terekhov, S.V. & Zakharov, D.N., (2001). Double-walled carbon nanotubes fabricated by a hydrogen arc discharge method. *Carbon,* 39, 761.

[15] Treacy, M. J., Ebbesen, T. W., & Gibson, J. M. (1996). Exceptionally high Young's modulus observed for individual carbon nanotubes. *Nature*, 381(6584), 678–680.

[16] Iijima, S., Brabec, Maiti, C., A., & Bernholc. J. (1996). Structural flexibility of carbon nanotubes. *Journal of Chemical Physics*, 104, 2089.

[17] Wong, E. W., Sheehan, P. E., & Lieber, C. M. (1997). Nanobeam mechanics: elasticity, strength, and toughness of nanorods and nanotubes. *Science*, 277(5334), 1971–1975.

[18] Falvo, M. R., Clary, G. J., Taylor II, R. M., Chi, V., Brooks Jr, F. P., Washburn, S., & Superfine, R. (1997). Bending and buckling of carbon nanotubes under large strain, *Nature*, 389, 532–534.

[19] Dujardin, E., Webbesen, T. W., Krishan, A., Yianilos, P. N., & Trecy, M. M. J. (1998). Young's modulus of single-walled nanotubes. *Phys. Rev. B*, 58, 14013.

[20] Bower, C., Rosen, R., Jin, L., Han, J., & Zhou. O. (1999). Deformation of carbon nanotubes in nanotube polymer composites. *Applied Physics Letters*, 74, 3317.

[21] Dekker. C., (1999). Carbon nanotubes as molecular quantum wires. *Physics Today*, 52, 22.

[22] Dillon, A., Jones, K. M., Bekkedahl, T. A., Kiang, C. H., Bethune, D. S., & Heben, M. J. (1997). Storage of hydrogen in single-walled carbon nanotubes. *Nature*, 386(6623), 377–379.

[23] Ren, Z., Lan, Y., Wang, Y. (2011). Aligned Carbon Nanotubes: Physics, Concepts, Fabrication and Devices. *Carbon Nanostructures*. Berlin: Springer.

[24] O'connell, M. J. (2006). Carbon nanotubes: properties and applications. CRC press.

[25] Georgantzinos, S. K., Giannopoulos, G. I. & Anifantis, N. K. (2009). An efficient numerical model for vibration analysis of single-walled carbon nanotubes*Computational Mechanics*, 43(6), 731–741.

[26] Dresselhaus, M. S., Dresselhaus, G., & Saito, R. (1995). Physics of carbon nanotubes.*Carbon*, 33(7), 883–891.

# 2

# Single-walled Carbon Nanotubes Modeled as Flügge Shell Theory: Influences of Length-to-Diameter Ratios

## Abstract

Based on Flügge shell theory, the vibrations of single-walled carbon nanotubes (SWCNTs) are investigated. Three categories of single-walled carbon nanotubes are used here: armchair, zigzag, and chiral. Influences of length-to-diameter ratios and the two boundary conditions on the natural frequencies of armchair, zigzag, and chiral SWCNTs are examined. The Rayleigh–Ritz method determines eigenfrequencies for single-walled carbon nanotubes. The solution is obtained using the geometric characteristics and boundary conditions for the natural frequencies of SWCNTs. The natural frequencies decrease as the length-to-diameter ratio increases, and the effect of frequencies is less significant and more prominent for a long tube. To assess the frequency, confirmation carried out in this chapter is compared with the earlier computations obtained from DTM and Bubnov–Galerkin method. Also, the results for three categories of SWCNTs are checked separately with corresponding results based on different models. Some results are also compared with experimental results of resonant Raman scattering (RRS). The present model is found to be fully validated by means of numerical and experimental results.

**Keywords:** Flügge shell theory, Rayleigh–Ritz method, armchair, zigzag and chiral SWCNTs, length-to-diameter ratios, boundary conditions.

## 2.1 Introduction

The structure of carbon nanotubes produced on the basis of a growth process. First, the production of carbon nanotubes appeared as ultra-thin graphene tubes with the cores of a hollow cylinder. Second, the growth of carbon nanotubes was observed as cylindrical carbon layers, which is hexagonally packed. The carbon nanotubes were originally reported by Iijima [1] . Later on, several material scientists [2, 3] probed the work of Iijima [1] and found it satisfactory. They found the use of carbon nanotubes in many fields like as material reinforcement, aerospace, medicine, defense, and microelectronic devices.

Treacy et al. [6] experimented MWCNTs to calculate the resonance frequency for clamped-free end conditions, which is associated with electrical loads or heat transfer. Peigl [7] developed a novel method by applying Euler–Bernoulli beam theory and nonuniform rational basis spline (NURBS) to model the waviness of aforesaid nano-scaled curved structures. Lordi and Yao [8] started the molecular dynamic simulation using interatomic potential function to develop a formula for approximating the tube radii. The vibration of chiral SWCNTs was calculated by Timoshenko beam model (TBM), which involves rotary inertia and shear deformation. The calculations that were performed with the cylindrical shell model differ slightly from those obtained with the MD simulation. The second method, which is the experimental method, was used by Krishnan et al. [9] to study the properties of CNTs. Jorio et al. [10] presented a review for isolated SWCNTs based on Raman spectra. Important structural information is given for single nanotube spectra. Many results of armchair, zigzag, and chiral SWCNTs are given in detail. Zhang et al. [11] investigated thin cylindrical shells' free vibrational behavior, engaging the Love shell equations. The finite element method has been used to present thin shell segmentation magnified with cohesive fracture. Li and Chou [12] used the molecular method for the vibrational behavior of CNTs and showed that the results of SWCNTs were 10% higher than those of DWCNTs of the same outer diameter. Pantano et al. [13] also used shell elements to effectively model and study the deformation of single and multiwall carbon nanotubes. Wang and Varadan [14] used nonlocal elastic shell theory to study the wave propagation analysis of carbon nanotubes. Wang and Zhang [15] examined the bending and torsional stiffness of single-walled CNT by applying the Flügge shell equations. They presented a three-dimensional model of single-walled CNT in their work with a thickness effect. Flügge shell theory (FST) was proposed to form governing equations of motion for CNTs. Natsuki et al. [16] predicted the vibrational characteristics of DWCNTs filled with fluid and governing equations derived from Flügge's shell equation for the vibration of CNTs. They investigated vibrational modes, parameter's influence, and CNT's fluid properties. Elishkoff and

Pentaras [17] applied the Petrov Galerkin and Dubnov–Galerkin methods to simulate the frequencies of CNTs and argued the explicit formula with different edge conditions. Lee and Chang [18] analyzed fluid-filled SWCNTs vibration mode shape and frequency using nonlocal elasticity theory. It is found that mode shape and frequency are influenced significantly by the nonlocal parameters, and the frequency component decreases as the nonlocal parameters increase. Gupta et al. [19] studied the free vibration of armchair, zigzag, and chiral SWCNTs to explore the aspect ratios using Rayleigh and Love inextensional vibration modes. The axial and torsional vibration modes are also exhibited. Simsek [20] used nonlocal EBM with supported edge conditions to find the forced vibration of SWCNTs. Effects are simulated with velocity and aspect ratio in detail. Moreover, the excitation frequency and load velocity are vital in the nanostructures. Yang et al. [21] demonstrated the frequencies of CNTs using nonlocal theory and geometric theory. The influence of nonlocal parameters on height and radius is studied in detail. Thai [22] also applied nonlocal theory to shear deformation theory across the thickness along with the vibration of shear strain. A series expansion was deployed to examine the buckling, fundamental frequencies, and deflection of CNTs. Rafiee et al. [23] conducted the CNT frequencies using EBM and the reinforced composite beam with piezoelectric CNTs based on Karman geometry. The effect of the surface concept with changing the temperature for linear and nonlinear CNTs has been studied comprehensively. Ansari and Arash [24] summarized the effect of small-scale, geometrical parameters and layer-wise end conditions of double-walled CNTs by adopting FST. Chawis et al. [25] investigated the effects of scale length and nonlocal parameters using Euler beam theory to determine the free vibration of single-walled CNTs. Ansari et al. [26] used the shell elements in a hybrid approach to model single-walled carbon nanotubes and compared it atomistic method. Ponnusamy and Amuthalakshmi [27] studied the vibrational characteristics of viscous fluid conveying double-walled carbon nanotube for clamped-clamped and clamped-free conditions. Strozzi et al. [28] proposed the low frequency of SWCNTs in the framework of Sanders Sanders–Koiter thin shell theory based on Rayleigh–Ritz method using clamped and free edge conditions. Different types of CNTs are considered in detail with various aspect ratios. Kiani [29] fully studied the vibration and instability of SWCNTs in a three-dimensional magnetic field. The bending frequency and its corresponding velocity were also reported. In addition, the buckling of CNTs occurs with varying magnetic fields. They have represented the nonlocal parameters that are highly influential on the bending vibration of a carbon nanotube. It was shown that the boundary conditions with nonlocal parameters are more effective on the nanotube vibration. Moreover, it has been observed that the frequencies decrease by increasing the nonlocal parameter on enhancing the length of CNT. Rouhi [30] studied the stability of SWCNTs under axial load and demonstrated the small-scale effects of lengths using nonlocal Flügge shell model based on Rayleigh–Ritz methods. It has been investigated that by adjusting nonlocal parameters, there is uncertainty in defining bending stiffness and in-plane stiffness. Rakrak [31] carried out the vibrational study of double-walled CNTs with the composition of Euler–Bernoulli beam and nonlocal elasticity theories. The dynamic aspect of double-walled CNTs has been explored by changing numerous physical and geometrical parameter of tubes. It has been observed that prominent C–C bond distortion was due to large curvature in short tubes. Ehyaei and Daman [32] investigated the vibration characteristics of SWCNTs and DWCNTs using the initial perfection and continuum mechanics approach. The general equation of motion was obtained by the Hamiltonian principle and energy equivalent model. The numerical frequencies of DWCNTs and SWCNTs were determined by Navier method and finite element method. Kumar [33] presented the vibration of double-walled carbon nanotubes with different boundary conditions using Euler–Bernoulli's elastic beams theory. To solve the governing equation, the differential transform method is utilized. The effect of Winkler elastic medium is also exhibited. Sobamowo et al. [34] studied the electrical and mechanical properties of single and double carbon nanotubes with a strength-to-width ratio based on nonlocal theory. They discussed many applications of single and double carbon nanotubes in physical and natural sciences. Miyashiro et al. [35] investigated the mechanical vibration of SWCNTs for different lengths with beam elements. It is found that the Bernoulli-Euler beam theory is a relationship between bending mode and the ratio of length-to-diameter. The modal analysis can be used accurately if the fundamental frequency is smaller than the fixed ratio. Jena et al. [36] devoted themselves to conducting the stability analysis of three types of single-walled carbon nanotubes based on Winkler foundation at low and high temperatures. The surface effects of armchair, zigzag, and chiral tubes are considered. Abdullah et al. [37] used nonlocal Euler–Bernoulli beam theory to investigate the vibration of cracked SWCNTs based on Winkler-type foundation using various boundary conditions. The influence of the scale coefficient in the elastic medium is conducted at low and high temperatures.

The simplicity of this model is due to its derived attractive results. Two sets of boundary conditions are adopted for the frequency behavior of three categories of SWCNTs According to these two boundary conditions, the nanotubes have clamped on both sides as one end condition, and for the other end condition, the nanotubes are clamped from one end, and the other end is simply supported. Excellent validation is observed with numerical and experimental results. The natural frequencies are affected by length-to-diameter ratios with two boundary conditions. This vibration investigation is the main basis for implementing SWCNTs in nanotechnology, like material reinforcement, aerospace, medicine, defense, and microelectronic devices.

## 2.2 Theoretical Formation

In past, for the modeling of carbon based nanostructure, many theories are presented. But some theories or models cannot capture the exact bending rigidity of CNTs with isotropic parameters [28] . But some models are more accurate with the selection of suitable parameters [38] . The structure of carbon nanotubes can be defined with material parameters as mass density $(\rho)$, Young's modulus (E), Poisson ratio $(v)$ and geometrical parameters as wall thickness $(h)$, length $(L)$, and diameter $(d)$. Figure 2.1 shows the graphene sheet with schematic representation of armchair, zigzag and chiral SWCNTs. When the graphene sheet is rolled then it looks like a hollow cylinder which termed as cylindrical shell (See Figure 2.2).

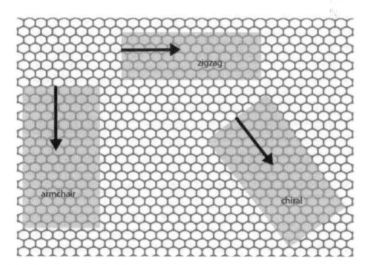

**Figure 2.1** Graphene sheet with a schematic representation of armchair, zigzag, and chiral SWCNTs.

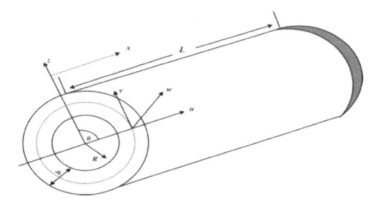

**Figure 2.2**    Definition of geometry of the tube, displacements, and stresses.

Figure 2.2 shows the rolled sheet of graphene with length $L$. The position of a point is represent with orthogonal coordinate system $(x, \theta, z)$ on the middle surface of shell which is treated as the axial $(x)$, circumferential $(\theta)$, and radial $(z)$ coordinate. However, these cylindrical coordinates can be used for resolving the shell energy expressions. This coordinate system has displacement deformations in the form as $u(x, \theta, t), v(x, \theta, t), w(x, \theta, t)$, where $t$ shows the configuration of the cross section.

### 2.2.1 Continuum Shell Theory

In the nineteenth century, English scientist Robert Hooke studied the relationships between stress–strain and concluded that many materials have similar formations. Here in the case of a cylindrical shell, the plane stress form is supposed and applied. This supposition is useful to convert the shell problem from 3D into 2D. According to 2D Hook's law, the mathematical form of the constitutive relations for a thin cylindrical shell (CS) are inscribed as

$$\{\sigma\} = [Q]\{e\} \tag{2.1}$$

where $\sigma$ and $e$ symbolizes the stress and strain vectors and $Q$ is associated with the reduced stiffness matrix.

If $\sigma_x, \sigma_\theta$ is the linear stresses, $e_x, e_\theta$ is the strains in the directions of $x$ and $\theta$ and $\sigma_{x\theta}, e_{x\theta}$ is the shear stress and strain, then the stress and strain column vectors takes the form

$$\{\sigma\}^T = \{\sigma_x, \sigma_\theta, \sigma_{x\theta}\}, \{e\}^T = \{e_x, e_\theta, e_{x\theta}\} \tag{2.2}$$

The plane stress supposition describes the mathematical formulation of the reduced stiffness matrix as

$$[Q] = \begin{pmatrix} Q_{11} & Q_{12} & 0 \\ Q_{12} & Q_{22} & 0 \\ 0 & 0 & Q_{66} \end{pmatrix} \tag{2.3}$$

Eqs. 2.2 and 2.3 are substituted in Eq. 2.1 to obtain:

$$\left\{ \begin{array}{c} \sigma_x \\ \sigma_\theta \\ \sigma_{x\theta} \end{array} \right\} = \begin{pmatrix} Q_{11} & Q_{12} & 0 \\ Q_{12} & Q_{22} & 0 \\ 0 & 0 & Q_{66} \end{pmatrix} \left\{ \begin{array}{c} e_x \\ e_\theta \\ e_{x\theta} \end{array} \right\} \tag{2.4}$$

For isotropic materials, the elements of $Q$ $[Q_{ij}(i,j = 1, 2, 6)]$ are defined as

$$Q_{11} = Q_{22} = \frac{E}{1 - v^2}, Q_{12} = \frac{vE}{1 - v^2}, Q_{66} = \frac{E}{2(1 + v)} \tag{2.5}$$

The linear combinations of thickness variable ($z$) and strain vectors components are developed as

$$e_x = e_1 + zk_1, e_\theta = e_2 + zk_2, e_{x\theta} = \gamma + 2z\tau \tag{2.6}$$

The references surface strains and curvatures are $(e_1, e_2, \gamma)$ and $(k_1, k_2, \tau)$

The components of stress vectors with the combination of eqn (2.4) and (2.6) can be written as

$$\sigma_x = (e_1 + zk_1) Q_{11} + (e_2 + zk_2) Q_{12}, \sigma_\theta = (e_1 + zk_1) Q_{12}$$
$$+ (e_2 + zk_2) Q_{22}, \sigma_{x\theta} = (\gamma + 2z\tau)Q_{66} \tag{2.7}$$

If $N_x (M_x), N_\theta (M_\theta), N_{x\theta} (M_{x\theta})$ are force (moment) components, respectively, in the directions of axial, circumferential and shear. The integral shapes of force and moments as under

$$\{N_x, N_\theta, N_{x\theta}\} = \int_{-\frac{h}{2}}^{\frac{h}{2}} \{\sigma_x, \sigma_\theta, \sigma_{x\theta}\} \, dz, \{M_x, M_\theta, M_{x\theta}\}$$

$$= \int_{-\frac{h}{2}}^{\frac{h}{2}} \{\sigma_x, \sigma_\theta, \sigma_{x\theta}\} \, zdz \tag{2.8}$$

The matrix is formed on combining the eqn (2.6) and (2.8)

$$\{N\} = [S]\{\varepsilon\} \tag{2.9}$$

where

$$\{N\}^T = \{N_x, N_\theta, N_{x\theta}, M_x, M_\theta, M_{x\theta}\}, \{\varepsilon\}^T = \{e_1, e_2, \gamma, k_1, k_2, 2\tau\},$$

$$[S] = \begin{pmatrix} A & B \\ B & D \end{pmatrix} \tag{2.10}$$

In matrix $S$, the elements $A$, $B$ and $D$ are the extensional, coupling and bending stiffness. The sub matrices of $A$, $B$ and $D$ with their entries as

$$[A] = \begin{pmatrix} A_{11} & A_{12} & 0 \\ A_{12} & A_{22} & 0 \\ 0 & 0 & A_{66} \end{pmatrix}, [B] = \begin{pmatrix} B_{11} & B_{12} & 0 \\ B_{12} & B_{22} & 0 \\ 0 & 0 & B_{66} \end{pmatrix},$$

$$[D] = \begin{pmatrix} D_{11} & D_{12} & 0 \\ D_{12} & D_{22} & 0 \\ 0 & 0 & D_{66} \end{pmatrix} \tag{2.11}$$

The complete form of [s] is

$$[S] = \begin{bmatrix} A_{11} & A_{12} & 0 & B_{11} & B_{12} & 0 \\ A_{12} & A_{22} & 0 & B_{12} & B_{22} & 0 \\ 0 & 0 & A_{66} & 0 & 0 & B_{66} \\ B_{11} & B_{12} & 0 & D_{11} & D_{12} & 0 \\ B_{12} & B_{22} & 0 & D_{12} & D_{22} & 0 \\ 0 & 0 & B_{66} & 0 & 0 & D_{66} \end{bmatrix} \tag{2.12}$$

where $A_y, B_{ij}$ and $D_{ij}$ $(i, j = 1, 2$ and $6)$ correspond to extensional, coupling and bending stiffness respectively and are written as:

$$A_{ij} = \int\limits_{-\frac{h}{2}}^{\frac{h}{2}} Q_{ij} dz, B_{ij} = \int\limits_{-\frac{h}{2}}^{\frac{h}{2}} Q_{ij} z dz, D_{ij} = \int\limits_{-\frac{h}{2}}^{\frac{h}{2}} Q_{ij} z^2 dz \tag{2.13}$$

The modified form eqn (2.9) can be attained with the help of eqn (2.10) and (2.12)

$$
\begin{bmatrix} N_r \\ N_\theta \\ N_{x\theta} \\ M_x \\ M_\theta \\ M_{x\theta} \end{bmatrix} = \begin{bmatrix} A_{11} & A_{12} & 0 & B_{11} & D_{12} & 0 \\ A_{12} & A_{22} & 0 & B_{12} & B_{22} & 0 \\ 0 & 0 & A_{66} & 0 & 0 & B_{66} \\ B_{11} & B_{12} & 0 & D_{11} & D_{12} & 0 \\ B_{12} & B_{22} & 0 & D_{12} & D_{22} & 0 \\ 0 & 0 & B_{66} & 0 & 0 & D_{66} \end{bmatrix} \begin{bmatrix} e_1 \\ e_2 \\ \gamma \\ k_1 \\ k_2 \\ 2\tau \end{bmatrix} \tag{2.14}
$$

In isotropic materials of cylindrical shell/tube, the coupling stiffness ceases to exist. The element in matrix for extensional and bending stiffness are given in Appendix 2.1.

The strain energy $U$ of vibrating shell/tube is shown below.

$$
U = \frac{1}{2} \int_0^L \int_0^{2\pi} \{\varepsilon\}^T [S] \{\varepsilon\} R d\theta dx \tag{2.15}
$$

Taking the value of $\{\varepsilon\}^T, [S]$ and $\{\varepsilon\}$ from Eqs. (2.10) and (2.12), the extended form strain energy of vibrating SWCNTs.

$$
U = \frac{1}{2} \int_0^L \int_0^{2\pi} \{ A_{11} e_1^2 + A_{22} e_2^2 + 2 A_{12} e_1 e_2 + A_{66} \gamma^2 + 2 B_{11} e_1 k_1 +
$$
$$
2 B_{12} e_1 k_2 + 2 B_{12} e_2 k_1 + 2 B_{22} e_2 k_2 + 4 B_{66} \gamma\tau + D_{11} k_1^2 + D_{22} k_2^2 +
$$
$$
2 D_{12} k_1 k_2 + 4 D_{65} \tau^2 \} R d\theta dx
$$
$$
\tag{2.16}
$$

Similarly, the kinetic energy $T$ of vibrating tube is given by:

$$
T = \frac{1}{2} \int_0^L \int_0^{2\pi} \rho_t \left[ \left( \frac{\partial u}{\partial t} \right)^2 + \left( \frac{\partial v}{\partial t} \right)^2 + \left( \frac{\partial w}{\partial t} \right)^2 \right] R d\theta dx \tag{2.17}
$$

where $\rho$ is the density and time variable is $t$.

$$
\rho_t = \int_{-\frac{h}{2}}^{\frac{h}{2}} \rho dz \tag{2.18}
$$

## 2.2.2 Flügge's shell theory

When the thickness of the equivalent shell is so small, then shell theory can capture the vibration of carbon nanotubes. Love Kirchhof's first approximations are used to derive the equations of Flügge's linear shell theory [39, 40] to investigate the vibrations of armchair, zigzag, and chiral single walled carbon nanotubes. The references surface strains $(e_1, e_2, \gamma)$ and curvatures $(k_1, k_2, \tau)$ are given as [39] .

$$\{e_1, e_2, e_3\} = \left\{ \frac{\partial u}{\partial x}, \frac{1}{R}\left( \frac{\partial v}{\partial \theta} + w \right), \frac{\partial v}{\partial x} + \frac{1}{R}\frac{\partial u}{\partial \theta} \right\} \qquad (2.19)$$

$$\{k_1, k_2, k_3\} = \left\{ -\frac{\partial^2 w}{\partial x^2}, -\frac{1}{R^2}\left( \frac{\partial^2 w}{\partial \theta^2} + w \right), -\frac{1}{R}\left( 2\frac{\partial^2 w}{\partial x \partial \theta} + \frac{1}{R}\frac{\partial u}{\partial \theta} - \frac{\partial v}{\partial x} \right) \right\} \qquad (2.20)$$

Eqn (2.19) and (2.20), are now combined in eqn (2.16) to attain a new form of strain energy $U$

$$
\begin{aligned}
U = \frac{1}{2} \int_0^L \int_0^{2\pi} &\left\{ A_{11}\left( \frac{\partial u}{\partial x} \right)^2 + \frac{A_{22}}{R^2}\left( \frac{\partial v}{\partial \theta} + w \right)^2 + \frac{2A_{12}}{R}\left( \frac{\partial u}{\partial x} \right)\left( \frac{\partial v}{\partial \theta} + w \right) \right. \\
&+ A_{66}\left( \frac{\partial v}{\partial x} + \frac{1}{R}\frac{\partial u}{\partial \theta} \right)^2 - 2B_{11}\left( \frac{\partial u}{\partial x} \right)\left( \frac{\partial^2 w}{\partial x^2} \right) - \frac{2B_{12}}{R^2}\left( \frac{\partial u}{\partial x} \right)\left( \frac{\partial^2 w}{\partial \theta^2} + w \right) \\
&- \frac{2B_{12}}{R}\left( \frac{\partial v}{\partial \theta} + w \right)\left( \frac{\partial^2 w}{\partial x^2} \right) - \frac{2B_{22}}{R^3}\left( \frac{\partial v}{\partial \theta} + w \right)\left( \frac{\partial^2 w}{\partial \theta^2} + w \right) \\
&- \frac{4B_{66}}{R}\left( \frac{\partial v}{\partial x} + \frac{1}{R}\frac{\partial u}{\partial \theta} \right)\left( 2\frac{\partial^2 w}{\partial x \partial \theta} + \frac{1}{R}\frac{\partial u}{\partial \theta} - \frac{\partial v}{\partial x} \right) + D_{11}\left( \frac{\partial^2 w}{\partial x^2} \right)^2 \\
&+ \frac{D_{22}}{R^4}\left( \frac{\partial^2 w}{\partial \theta^2} + w \right)^2 + \frac{2D_{12}}{R^2}\left( \frac{\partial^2 w}{\partial x^2} \right)\left( \frac{\partial^2 w}{\partial \theta^2} + w \right) \\
&\left. + \frac{4D_{66}}{R^2}\left( 2\frac{\partial^2 w}{\partial x \partial \theta} + \frac{1}{R}\frac{\partial u}{\partial \theta} - \frac{\partial v}{\partial x} \right)^2 \right\} R\,d\theta\,dx
\end{aligned}
$$

$$(2.21)$$

The Lagrangian expression $\Pi$ is developed after establishing the expressions of strain and kinetic energy

$$\Pi = T - U \qquad (2.22)$$

### 2.2.3 Solution scheme

In general formulations for the problems of shell/tube/plates with different end conditions [41], the analytical or exact solution is impossible or complex, which is not valid for practical purposes. The solution of a mathematical problem is done by numerical method. The original problem is reduced by repeating the same steps or series. In the past, some authors used the Donnell equation, finite element method, and variational principles [42]. The frequency equations are extracted with the Rayleigh–Ritz method in the present model. The equation of motion of shells/tubes is converted with axial modal dependence. To measure the axial modal dependence, many functions (spline, polynomial, beam, and Fourier series forms) are used to satisfy the geometric conditions.

### 2.2.4 Selection of displacement deformation functions

After developing the strain and kinetic energy expressions, the upcoming step is to select the suitable displacement functions that are essential in the Rayleigh–Ritz procedure. The displacement deformation functions $u(x, \theta, t)$, $v(x, \theta, t)$, $w(x, \theta, t)$ in the directions of axial, circumferential, and radial. Generally, for shell/tube problems, the displacement functions are articulated by separating variables to satisfy the edge conditions. The space and time variables are separated. This method of separation is widely used for solving PDEs. By the Rayleigh–Ritz procedure, a system of homogeneous equations with three vibration amplitudes coefficients is attained after substituting the assumed displacement functions into the shell equation of motion. The assumed modal displacement deformation functions are selected as under [43]

$$u(x, \theta, t) = A_m \frac{\partial \varphi}{\partial x} \sin n\theta \sin \omega t \qquad (2.23a)$$

$$v(x, \theta, t) = B_m \varphi(x) \cos n\theta \sin \omega t \qquad (2.23b)$$

$$w(x, \theta, t) = C_m \varphi(x) \sin n\theta \sin \omega t \qquad (2.23c)$$

where $\omega, n$ in Eq. (2.23a–2.23c) is the natural frequency and circumferential wave number. The coefficients $A_m, B_m, C_m$ the amplitudes of the vibrations in the $x, \theta$, and $z$ directions respectively. $\varphi(x)$ is the axial function. Substitute the Eq. (2.23a,b,c) along with their partial derivatives in Eqs. (2.17–2.21). The obtained equations are as under:

$$
T = \frac{\omega^2 R \rho t}{2} \left[ \int_0^L \int_0^{2\pi} \left( A_m^2 \left( \frac{d\phi}{dx} \right)^2 \sin^2 n\theta + B_m^2 \varphi^2(x) \cos^2 n\theta \right. \right.
$$
$$
\left. \left. + C_m^2 \phi^2(x) \sin^2 n\theta \right) d\theta dx \right] \cos^2 \omega t \tag{2.24}
$$

$$
U = \frac{R}{2} \int_0^L \int_0^{2\pi} \left[ \left( A_{11} A_m^2 \left( \frac{d^2\varphi}{dx^2} \right)^2 \right) \sin^2 n\theta + \frac{A_{22}}{R^2} \left( n^2 B_m^2 \varphi^2(x) + \right. \right.
$$

$$
C_m^2 \varphi^2(x) - 2n B_m C_m \varphi^2(x) \right) \sin^2 n\theta + \frac{2A_{12}}{R} \left( -n A_m B_m \left( \frac{d^2\varphi}{dx^2} \right) \varphi(x) \right.
$$

$$
\left. + A_m C_m \left( \frac{d^2\varphi}{dx^2} \right) \varphi(x) \right) \sin^2 n\theta + A_{66} \left( B_m^2 \left( \frac{d\varphi}{dx} \right)^2 + \frac{n^2 A_m^2}{R^2} \left( \frac{d\varphi}{dx} \right)^2 + \right.
$$

$$
\frac{2n}{R} A_m B_m \left( \frac{d\varphi}{dx} \right)^2 \right) \cos^2 n\theta - 2B_{11} \left( A_m C_m \left( \left( \frac{d^2\varphi}{dx^2} \right)^2 \right) \right) \sin^2 n\theta
$$

$$
-2\frac{B_{12}}{R^2} \left( -n^2 A_m C_m \left( \frac{d^2\varphi}{dx^2} \right) \varphi(x) + A_m C_m \left( \frac{d^2\varphi}{dx^2} \right) \varphi(x) \right) \sin^2 n\theta
$$

$$
-\frac{2B_{12}}{R} \left( -n B_m C_m \frac{d^2\varphi}{dx^2} \varphi(x) + C_m^2 \frac{d^2\varphi}{dx^2} \varphi(x) \right) \sin^2 n\theta
$$

$$
-\frac{2B_{22}}{R^3} \left( n^3 B_m C_m \varphi^2(x) - n B_m C_m \varphi^2(x) - n^2 C_m^2 \varphi^2(x) \right.
$$

$$
\left. + C_m^2 \varphi^2(x) \right) \sin^2 n\theta + \frac{4B_{66}}{R}
$$

$$
\left[ \begin{array}{l} 2n B_m C_m \left( \frac{\partial\varphi}{\partial x} \right)^2 + \frac{n}{R} A_m B_m \left( \frac{\partial\varphi}{\partial x} \right)^2 - B_m^2 \left( \frac{\partial\varphi}{\partial x} \right)^2 \\ + \frac{n^2}{R} A_m C_m \left( \frac{\partial\varphi}{\partial x} \right)^2 + \frac{n^2}{R} A_m^2 \left( \frac{\partial\varphi}{\partial x} \right)^2 - \frac{n}{R} A_m B_m \left( \frac{\partial\varphi}{\partial x} \right)^2 \end{array} \right] \cos^2 n\theta
$$

$$
+ D_{11} C_m^2 \left( \frac{\partial^2\varphi}{\partial x^2} \right)^2 \sin^2 n\theta + \frac{D_2}{R^4} \left( n^4 C_m^2 \varphi^2(x) + C_m^2 \varphi^2(x) \right.
$$

$$
\left. -2n^2 C_m^2 \varphi^2(x) \right) \sin^2 n\theta + \frac{2D_{12}}{R^2} \left( -n^2 C_m^2 \frac{d^2\varphi}{dx^2} \varphi(x) + C_m^2 \frac{d^2\varphi}{dx^2} \varphi(x) \right) \sin^2 n\theta
$$

2.2 *Theoretical Formation* 23

$$+\frac{4D_{66}}{R^2}\left(\begin{array}{l}4n^2C_m^2\left(\frac{\partial\varphi}{\partial x}\right)^2+\frac{n^2}{R^2}A_m^2\left(\frac{\partial\varphi}{\partial x}\right)^2+B_m^2\left(\frac{\partial\varphi}{\partial x}\right)^2\\[2mm]+\frac{4n^2}{R}A_mC_m\left(\frac{\partial\varphi}{\partial x}\right)^2\\[2mm]-2\frac{n}{R}A_mB_m\left(\frac{\partial\varphi}{\partial x}\right)^2-4nB_mC_m\left(\frac{\partial\varphi}{\partial x}\right)^2\end{array}\right)\cos^2 n\theta\Bigg]\sin^2\omega t$$

$$(2.25)$$

Substitute the above expressions of the shell energies into eqn (2.22), the new expression for the Lagrangian functional is achieved as

$$\Pi=\frac{\omega^2 R\rho t}{2}\left[\int_0^L\int_0^{2\pi}\left(A_m^2\left(\frac{d\varphi}{dx}\right)^2\sin^2 n\theta+B_m^2\varphi^2(x)\cos^2 n\theta\right.\right.$$

$$\left.+C_m^2\varphi^2(x)\sin^2 n\theta\right)d\theta dx\Bigg]\cos^2\omega t-\frac{R}{2}\int_0^L\int_0^{2\pi}\left[\left(A_{11}A_m^2\left(\frac{d^2\varphi}{dx^2}\right)^2\right)\right.$$

$$\sin^2 n\theta+\frac{A_{22}}{R^2}\left(n^2B_m^2\varphi^2(x)+C_m^2\varphi^2(x)-2nB_mC_m\varphi^2(x)\right)\sin^2 n\theta$$

$$+\frac{2A_{12}}{R}\left(-nA_mB_m\left(\frac{d^2\varphi}{dx^2}\right)\varphi(x)+A_mC_m\left(\frac{d^2\varphi}{dx^2}\right)\varphi(x)\right)\sin^2 n\theta$$

$$+A_{66}\left(B_m^2\left(\frac{d\varphi}{dx}\right)^2+\frac{n^2A_m^2}{R^2}\left(\frac{d\varphi}{dx}\right)^2+\frac{2n}{R}A_mB_m\left(\frac{d\varphi}{dx}\right)^2\right)\cos^2 n\theta-2$$

$$B_{11}\left(A_mC_m\left(\left(\frac{d^2\varphi}{dx^2}\right)^2\right)\right)\sin^2 n\theta-2\frac{B_{12}}{R^2}\left(-n^2A_mC_m\left(\frac{d^2\varphi}{dx^2}\right)\varphi(x)\right.$$

$$+A_mC_m\left(\frac{d^2\varphi}{dx^2}\right)\varphi(x)\right)\sin^2 n\theta-\frac{2B_{12}}{R}\left(-nB_mC_m\frac{d^2\varphi}{dx^2}\right.$$

$$\varphi(x)+C_m^2\frac{d^2\varphi}{dx^2}\varphi(x)\right)\sin^2 n\theta-\frac{2B_{22}}{R^3}$$

$$\left(n^3B_mC_m\varphi^2(x)-nB_mC_m\varphi^2(x)-n^2C_m^2\varphi^2(x)+C_m^2\varphi^2(x)\right)\sin^2 n\theta$$

$$+\frac{4B_{66}}{R}\left[\begin{array}{l}2nB_mC_m\left(\frac{\partial\varphi}{\partial x}\right)^2+\frac{n}{R}A_mB_m\left(\frac{\partial\varphi}{\partial x}\right)^2-B_m^2\left(\frac{\partial\varphi}{\partial x}\right)^2\\[2mm]+\frac{n^2}{R}A_mC_m\left(\frac{\partial\varphi}{\partial x}\right)^2+\frac{n^2}{R}A_m^2\left(\frac{\partial\varphi}{\partial x}\right)^2-\frac{n}{R}A_mB_m\left(\frac{\partial\varphi}{\partial x}\right)^2\end{array}\right]\cos^2 n\theta$$

$$+ D_{11} C_m^2 \left( \frac{\partial^2 \varphi}{\partial x^2} \right)^2 \sin^2 n\theta + \frac{D_2}{R^4} \left( n^4 C_m^2 \varphi^2(x) + C_m^2 \varphi^2(x) - 2n^2 C_m^2 \varphi^2(x) \right)$$

$$\sin^2 n\theta + \frac{2D_{12}}{R^2} \left( -n^2 C_m^2 \frac{d^2\varphi}{dx^2} \varphi(x) + C_m^2 \frac{d^2\varphi}{dx^2} \varphi(x) \right) \sin^2 n\theta$$

$$+ \frac{4D_{66}}{R^2} \left[ \begin{array}{c} 4n^2 C_m^2 \left( \frac{\partial \varphi}{\partial x} \right)^2 + \frac{n^2}{R^2} A_m^2 \left( \frac{\partial \varphi}{\partial x} \right)^2 + B_m^2 \left( \frac{\partial \varphi}{\partial x} \right)^2 \\ + \frac{4n^2}{R} A_m C_m \left( \frac{\partial \varphi}{\partial x} \right)^2 \\ - 2\frac{n}{R} A_m B_m \left( \frac{\partial \varphi}{\partial x} \right)^2 - 4n B_m C_m \left( \frac{\partial \varphi}{\partial x} \right)^2 \end{array} \right] \cos^2 n\theta \Bigg] \sin^2 \omega t$$

$$(2.26)$$

## 2.2.5 Derivation of generalized eigenvalue problem

The Rayleigh–Ritz method is employed to investigate the vibration characteristics of the present shell problem. This method is based on the principle
of minimization of energy. Now the minimum kinetic and strain energies of
the cylindrical shell are written as

$$T_{\max} = \frac{\pi \omega^2 R \rho t}{2} \left[ \int_0^L \left( A_m^2 \left( \frac{d\varphi}{dx} \right)^2 + B_m^2 \varphi^2(x) + C_m^2 \varphi^2(x) \right) dx \right]$$

$$(2.27)$$

$$U_{\max} = \frac{\pi R}{2} \int_0^L \left[ A_{11} A_m^2 \left( \frac{d^2\varphi}{dx^2} \right)^2 + \frac{A_{22}}{R^2} \left( n^2 B_m^2 \varphi^2(x) + C_m^2 \varphi^2(x) \right) \right.$$

$$-2n B_m C_m \varphi^2(x)) + \frac{2A_{12}}{R} \left( -n A_m B_m \left( \frac{d^2\varphi}{dx^2} \right) \varphi(x) + A_m C_m \left( \frac{d^2\varphi}{dx^2} \right) \right.$$

$$\varphi(x)) + A_{66} \left( B_m^2 \left( \frac{d\varphi}{dx} \right)^2 + \frac{n^2 A_m^2}{R^2} \left( \frac{d\varphi}{dx} \right)^2 + \frac{2n}{R} A_m B_m \left( \frac{d\varphi}{dx} \right)^2 \right)$$

$$-2B_{11} \left( A_m C_m \left( \left( \frac{d^2\varphi}{dx^2} \right)^2 \right) \right) \sin^2 n\theta - 2\frac{B_{12}}{R^2} \left( -n^2 A_m C_m \left( \frac{d^2\varphi}{dx^2} \right) \varphi(x) \right.$$

$$+ A_m C_m \left( \frac{d^2\varphi}{dx^2} \right) \varphi(x)) - \frac{2B_{12}}{R} \left( -n B_m C_m \frac{d^2\varphi}{dx^2} \varphi(x) + C_m^2 \frac{d^2\varphi}{dx^2} \varphi(x) \right)$$

$$- \frac{2B_{22}}{R^3} \left( n^3 B_m C_m \varphi^2(x) - n B_m C_m \varphi^2(x) - n^2 C_m^2 \varphi^2(x) + C_m^2 \varphi^2(x) \right)$$

$$
+ \frac{4B_{66}}{R} \left[ \begin{array}{l} 2n D_m C_m \left(\frac{\partial \varphi}{\partial x}\right)^2 + \frac{n}{R} A_m B_m \left(\frac{\partial \psi}{\partial x}\right)^2 - B_m^2 \left(\frac{\partial \varphi}{\partial x}\right)^2 \\ + \frac{n^2}{R} A_m C_m \left(\frac{\partial \varphi}{\partial x}\right)^2 + \frac{n^2}{R} A_m^2 \left(\frac{\partial \varphi}{\partial x}\right)^2 - \frac{n}{R} A_m B_m \left(\frac{\partial \varphi}{\partial x}\right)^2 \end{array} \right]
$$

$$
+ D_{11} C_m^2 \left(\frac{\partial^2 \varphi}{\partial x^2}\right)^2 \sin^2 n\theta + \frac{D_m}{R^4} \left( n^4 C_m^2 \varphi^2(x) + C_m^2 \varphi^2(x) - 2n^2 C_m^2 \varphi^2(x) \right)
$$

$$
+ \frac{2D_{12}}{R^2} \left( -n^2 C_m^2 \frac{d^2 \varphi}{dx^2} \varphi(x) + C_m^2 \frac{d^2 \varphi}{dx^2} \varphi(x) \right)
$$

$$
+ \frac{4D_{66}}{R^2} \left[ \begin{array}{c} 4n^2 C_m^2 \left(\frac{\partial \varphi}{\partial x}\right)^2 + \frac{n^2}{R^2} A_m^2 \left(\frac{\partial \varphi}{\partial x}\right)^2 + B_m^2 \left(\frac{\partial \varphi}{\partial x}\right)^2 \\ + \frac{4n^2}{R} A_m C_m \left(\frac{\partial \varphi}{\partial x}\right)^2 \\ -2\frac{n}{R} A_m B_m \left(\frac{\partial \varphi}{\partial x}\right)^2 - 4n B_m C_m \left(\frac{\partial \varphi}{\partial x}\right)^2 \end{array} \right] dx \qquad (2.28)
$$

by taking $\sin^2 \omega t = \cos^2 \omega t = 1$ and $\int_0^L \sin^2 n\theta d\theta = \int_0^L \cos^2 n\theta d\theta = \pi$

Hence

$$
\Pi = T_{\max} - U_{\max} \qquad (2.29)
$$

Making substitutions of expressions for $T_{\max}$, and $U_{\max}$ the following form of the Lagrangian functional is written as:

$$
\Pi = \frac{\pi R}{2} \left[ \frac{\omega^2 \rho t}{2} \int_0^L \left( A_m^2 \left(\frac{d\varphi}{dx}\right)^2 + B_m^2 \varphi^2(x) + C_m^2 \varphi^2(x) \right) dx - \right.
$$

$$
\int_0^L \left[ A_{11} A_m^2 \left(\frac{d^2 \varphi}{dx^2}\right)^2 + \frac{A_{22}}{R^2} \left( n^2 B_m^2 \varphi^2(x) + C_m^2 \varphi^2(x) - 2n B_m C_m \varphi^2(x) \right) \right.
$$

$$
+ \frac{2A_{12}}{R} \left( -n A_m B_m \left(\frac{d^2 \varphi}{dx^2}\right) \varphi(x) + A_m C_m \left(\frac{d^2 \varphi}{dx^2}\right) \varphi(x) \right) +
$$

$$
A_{66} \left( B_m^2 \left(\frac{d\varphi}{dx}\right)^2 + \frac{n^2 A_m^2}{R^2} \left(\frac{d\varphi}{dx}\right)^2 + \frac{2n}{R} A_m B_m \left(\frac{d\varphi}{dx}\right)^2 \right) \right)
$$

$$
- 2B_{11} \left( A_m C_m \left( \left(\frac{d^2 \varphi}{dx^2}\right)^2 \right) \sin^2 n\theta - 2\frac{B_{12}}{R^2} \left( -n^2 A_m C_m \left(\frac{d^2 \varphi}{dx^2}\right) \varphi(x) \right.
$$

$$+A_m C_m \left( \frac{d^2\varphi}{dx^2} \right) \varphi(x) \Bigg)$$

$$- \frac{2B_{12}}{R} \left( -nB_m C_m \frac{d^2\varphi}{dx^2} \varphi(x) + C_m^2 \frac{d^2\varphi}{dx^2} \varphi(x) \right)$$

$$- \frac{2B_{22}}{R^3} \left( n^3 B_m C_m \varphi^2(x) - nB_m C_m \varphi^2(x) - n^2 C_m^2 \varphi^2(x) + C_m^2 \varphi^2(x) \right)$$

$$+ \frac{4B_{66}}{R} \left[ \begin{matrix} 2nB_m C_m \left( \frac{\partial\varphi}{\partial x} \right)^2 + \frac{n}{R} A_m B_m \left( \frac{\partial\varphi}{\partial x} \right)^2 \\ -B_m^2 \left( \frac{\partial\varphi}{\partial x} \right)^2 + \frac{n^2}{R} A_m C_m \left( \frac{\partial\varphi}{\partial x} \right)^2 \\ + \frac{n^2}{R} A_m^2 \left( \frac{\partial\varphi}{\partial x} \right)^2 - \frac{n}{R} A_m B_m \left( \frac{\partial\varphi}{\partial x} \right)^2 \end{matrix} \right]$$

$$+ D_{11} C_m^2 \left( \frac{\partial^2\varphi}{\partial x^2} \right)^2 \sin^2 n\theta + \frac{D_{m2}}{R^4} \left( n^4 C_m^2 \varphi^2(x) + C^2 \varphi^2(x) - 2n^2 C_m^2 \varphi^2(x) \right)$$

$$+ \frac{2D_{12}}{R^2} \left( -n^2 C_m^2 \frac{d^2\varphi}{dx^2} \varphi(x) + C_m^2 \frac{d^2\varphi}{dx^2} \varphi(x) \right)$$

$$+ \frac{4D_{66}}{R^2} \left[ \begin{matrix} 4n^2 C_m^2 \left( \frac{\partial\varphi}{\partial x} \right)^2 + \frac{n^2}{R^2} A_m^2 \left( \frac{\partial\varphi}{\partial x} \right)^2 + B_m^2 \left( \frac{\partial\varphi}{\partial x} \right)^2 \\ + \frac{4n^2}{R} A_m C_m \left( \frac{\partial\varphi}{\partial x} \right)^2 \\ -2\frac{n}{R} A_m B_m \left( \frac{\partial\varphi}{\partial x} \right)^2 - 4nB_m C_m \left( \frac{\partial\varphi}{\partial x} \right)^2 \end{matrix} \right]$$

$$\text{(2.30)}$$

Applying the Rayleigh-Ritz procedure, $\Pi$ is minimized with respect to the vibration amplitudes $A_m$, $B_m$, and $C_m$, So extremizing $\Pi$, the following required conditions are obtained:

$$\frac{\partial\Pi}{\partial A_m} = \frac{\partial\Pi}{\partial B_m} = \frac{\partial\Pi}{\partial C_m} = 0 \qquad \text{(2.31)}$$

where

$$\frac{\partial\Pi}{\partial A_m} = \pi R \left[ \omega^2 \rho t A_m \int_0^L \left( \frac{d\varphi}{dx} \right)^2 dx \right.$$

$$\left. - \left( \begin{matrix} A_{11} \int_0^L \left( \frac{d^2\varphi}{dx^2} \right)^2 dx + \frac{n^2 A_{66}}{R^2} \int_0^L \left( \frac{d\varphi}{dx} \right)^2 dx \\ + \frac{4n^2 B_{66}}{R^2} \int_0^L \left( \frac{d\varphi}{dx} \right)^2 dx + \frac{4n^2 D_{66}}{R^4} \int_0^L \left( \frac{d\varphi}{dx} \right)^2 dx \end{matrix} \right) A_m \right]$$

$$+ \left( -\frac{2nA_{12}}{R} \int_0^L \left( \frac{d^2\varphi}{dx^2} \right) \varphi(x)dx + \frac{2nA_{66}}{R} \int_0^L \left( \frac{d\varphi}{dx} \right)^2 dx \right.$$

$$\left. -\frac{8nD_{66}}{R^3} \int_0^L \left( \frac{d\varphi}{dx} \right)^2 dx \right) B_m$$

$$+ \left( \begin{array}{c} \frac{2A_{12}}{R} \int_0^L \left( \frac{d^2\varphi}{dx^2} \right) \varphi(x)dx - 2B_{11} \int_0^L \left( \frac{d^2\varphi}{dx^2} \right)^2 dx \\[2mm] -\frac{2B_{12}}{R^2} \left( -n^2 \int_0^L \left( \frac{d^2\varphi}{dx^2} \right) \varphi(x)dx + \int_0^L \left( \frac{d^2\varphi}{dx^2} \right) \varphi(x)dx \right) \\[2mm] +\frac{4n^2 B_{66}}{R^2} \int_0^L \left( \frac{d\varphi}{dx} \right)^2 dx + \frac{16n^2 D_{66}}{R^3} \int_0^L \left( \frac{d\varphi}{dx} \right)^2 dx \end{array} \right) C_m \Bigg]$$

$$(2.32a)$$

$$\frac{\partial \Pi}{\partial B_m} = \pi R \left[ \omega^2 \rho_t B_m \int_0^L \varphi^2(x)dx \right.$$

$$- \left( -\frac{2nA_{12}}{R} \int_0^L \left( \frac{d^2\varphi}{dx^2} \right) \varphi(x)dx + \frac{2nA_{66}}{R} \int_0^L \left( \frac{d\varphi}{dx} \right)^2 dx \right.$$

$$\left. -\frac{8nD_{66}}{R^3} \int_0^L \left( \frac{d\varphi}{dx} \right)^2 dx \right) A_m$$

$$+ \left( \frac{n^2 A_{22}}{R^2} \int_0^x \varphi^2(x)dx + \frac{A_{66}}{R} \int_0^L \left( \frac{d\varphi}{dx} \right)^2 dx \right. \qquad (2.32b)$$

$$\left. -\frac{4B_{66}}{R} \int_0^L \left( \frac{d\varphi}{dx} \right)^2 dx + \frac{4D_{66}}{R^2} \int_0^L \left( \frac{d\varphi}{dx} \right)^2 dx \right) B_m +$$

$$\left( \begin{array}{c} -\frac{2nA_{22}}{R^2} \int_0^L \varphi^2(x)dx + \frac{2nB_{12}}{R} \int_0^L \frac{d^2\varphi}{dx^2} \varphi(x)dx \\[2mm] -\frac{2B_{22}}{R^3} \left( n^3 \int_0^L \varphi^2(x) - n \int_0^L \varphi^2(x) \right) \\[2mm] +\frac{8nB_{66}}{R} \int_0^L \left( \frac{d\varphi}{dx} \right)^2 dx + \frac{16nD_{66}}{R^2} \int_0^L \left( \frac{d\varphi}{dx} \right)^2 dx \end{array} \right) C_m \Bigg]$$

$$\frac{\partial \Pi}{\partial C_m} = \pi R \left\{ \omega^2 \rho_t C_m \int_0^L \varphi^2(x) dx - \left( \frac{2A_{12}}{R} \int_0^L \left( \frac{d^2\varphi}{dx^2} \right) \varphi(x) dx \right. \right.$$

$$- 2B_{11} \int_0^L \left( \frac{d^2\varphi}{dx^2} \right)^2 dx - \frac{2B_{12}}{R^2} \left( -n^2 \int_0^L \left( \frac{d^2\varphi}{dx^2} \right) \varphi(x) dx \right.$$

$$\left. + \int_0^L \left( \frac{d^2\varphi}{dx^2} \right) \varphi(x) dx \right) + \frac{4n^2 B_{66}}{R^2} \int_0^L \left( \frac{d\varphi}{dx} \right)^2 dx$$

$$\left. \left. + \frac{16n^2 D_{66}}{R^3} \int_0^L \left( \frac{d\varphi}{dx} \right)^2 dx \right) \right\} A_m$$

$$+ \left( \begin{array}{c} -\frac{2nA_{22}}{R^2} \int_0^L \phi^2(x) dx + \frac{2nB_{12}}{R} \int_0^L \frac{d^2\varphi}{dx^2} \varphi(x) dx \\[2mm] -\frac{2B_{22}}{R^3} \left( n^3 \int_0^L \varphi^2(x) - n \int_0^L \varphi^2(x) \right) + \frac{8nB_{66}}{R} \int_0^L \left( \frac{d\varphi}{dx} \right)^2 dx \\[2mm] +\frac{16nD_{66}}{R^2} \int_0^L \left( \frac{d\varphi}{dx} \right)^2 dx \end{array} \right) B_m$$

$$+ \left( \begin{array}{c} \frac{A_{22}}{R^2} \int_0^L \varphi^2(x) dx - \frac{2B_{12}}{R} \int_0^L \frac{d^2\varphi}{dx^2} \varphi(x) dx - \frac{2B_{22}}{R^3} \left( -n^2 \int_0^L \varphi^2(x) dx + \int_0^L \varphi^2(x) dx \right) + \\[2mm] D_{11} \int_0^L \left( \frac{d^2\varphi}{dx^2} \right)^2 dx + \frac{D_{22}}{R^4} \left( n^4 \int_0^L \varphi^2(x) dx + \int_0^L \varphi^2(x) dx - 2n^2 \int_0^L \varphi^2(x) dx \right) \\[2mm] +\frac{2D_{12}}{R^2} \left( -n^2 \int_0^L \frac{d^2\varphi}{dx^2} \varphi(x) dx + \int_0^L \frac{d^2\varphi}{dx^2} \varphi(x) dx \right) + \frac{16n^2 D_{66}}{R^2} \int_0^L \left( \frac{d^2\varphi}{dx^2} \right)^2 dx \end{array} \right) C_m$$

$$(2.33c)$$

## 2.2.6 Derivation of shell frequency equation

Shell frequency equation is obtained from the extrema conditions (2.32a) to (2.32c). After rearranging the terms in these conditions, the frequency equation is achieved in the following eigenvalue problem

$$C_{11}A_m + C_{12}B_m + C_{13}C_m = \omega^2 \rho_t A_m \qquad (2.32a)$$

$$C_{21}A_m + C_{22}B_m + C_{23}C_m = \omega^2 \rho_t B_m \qquad (2.33b)$$

$$C_{31}A_m + C_{32}B_m + C_{33}C_m = \omega^2 \rho_t C_m \qquad (2.34c)$$

where $C_i$ 's $(i, j = 1, 2, 3)$ are defined in Appendix 2.2.

The Equ, (34) can be written in the matrix notation as:

$$\begin{pmatrix} C_{11} & C_{12} & C_{13} \\ C_{21} & C_{22} & C_{23} \\ C_{31} & C_{32} & C_{33} \end{pmatrix} \begin{Bmatrix} A_m \\ B_m \\ C_m \end{Bmatrix} = \omega^2 \rho_t \begin{pmatrix} 1 & 0 & 0 \\ 0 & 1 & 0 \\ 0 & 0 & 1 \end{pmatrix} \begin{Bmatrix} A_m \\ B_m \\ C_m \end{Bmatrix} \quad (2.35)$$

For getting the shell frequency equation, the determinant of the matrix coefficients has vanished for the nontrivial eigenvalue of the shell frequency. Then, either the eigenvalue problem obtained above is solved by using computer program or the determinant is extended and the frequencies equation is obtained in the way of a polynomial equation involving $\omega^2$

## 2.3 Result and Discussion

In this section, the natural frequencies of armchair, zigzag, and chiral SWCNTs with two boundary conditions, namely, clamped-clamped (C–C) and clamped-simply supported (C–SS), modeled as Flügge's shell theory based on Rayleigh–Ritz method. In order to investigate the influence of natural frequencies versus length-to-diameter ratio, one needs to choose the geometrical and physical parameters. The successful application of the present model is to select the appropriate parameters. The thickness of CNTs is taken as $h = 0.34$ nm. The ratio of Young's modulus and mass density ($E/\rho = 3.6481 \times 10^8$ m$^2$/s$^2$) is taken from MD simulation. The stiffness ($Eh = 278.25$ GPa·nm) is calculated from ratio $E/\rho$, and Poisson's ratio is $\nu = 0.2$. These parameters are taken from the study of Zhang et al. [44] . For validating the present study, the results are presented in Tables 2.1–2.4. Table 2.1 predicted the comparison of the present model with the results reported by Kumar [33] and Elishakoff and Pentaras [17] . Kumar [33] and Elishakoff and Pentaras [17] conducted the vibration of carbon nanotubes with DTM and Bubnov–Galerkin method (BGM) [17] . The results of these two different models are much closed to the present model, which specify the validation of this study. The armchair and zigzag natural frequencies in Tables 2.2 and 2.3 are well-matched with the results of Strozzi et al. [28] and Gupta et al. [19] .

Table 2.4 shows the chiral frequency comparison with Strozzi et al. [28] and Jorio et al. [10] . The analytic and numerical methods used by Kumar [33], Elishakoff and Pentaras [17], Strozzi et al. [28], Gupta et al. [19], and Jorio et al. [10] investigated the experimental results of carbon nanotubes. So the natural frequencies computed by the present model are validated by analytical, numerical, and experimental results, showing a high convergence

**Table 2.1** Comparison of RRM frequencies of SWCNTs with DTM [33] and BGM [17] .

| L/d | Frequency (THz) | | |
|-----|-----------------|----------------|-----------------------------|
| | Present | DTM Kumer [33] | BGM Elishakoff and Pentaras [17] |
| 10 | 1.06387 | 1.06406 | 1.07986 |
| 12 | 0.73572 | 0.73683 | 0.75063 |
| 14 | 0.54107 | 0.54256 | 0.55171 |
| 16 | 0.41371 | 0.41371 | 0.42248 |
| 18 | 0.32592 | 0.32654 | 0.33385 |
| 20 | 0.26516 | 0.26546 | 0.27043 |

**Table 2.2** Comparison of RRM frequencies of armchair SWCNTs with Strozzi et al. [28] and Gupta et al. [19] .

| (m, n) | Frequency (THz) | | |
|----------|-----------------|-----------------|-----------------|
| | Present | Strozzi et al. [28] | Gupta et al. [19] |
| (6, 6) | 8.567 | 8.636 | 8.348 |
| (7, 7) | 7.287 | 7.399 | 7.166 |
| (8, 8) | 6.399 | 6.473 | 6.275 |
| (10, 10) | 5.117 | 5.184 | 5.026 |
| (11, 11) | 4.675 | 4.711 | 4.917 |
| (12, 12) | 4.287 | 4.318 | 4.19 |
| (15, 15) | 3.412 | 3.453 | 3.354 |
| (18, 18) | 2.825 | 2.878 | 2.796 |
| (20, 20) | 2.545 | 2.590 | 2.516 |

**Table 2.3** Comparison of RRM frequencies of zigzag SWCNTs with Strozzi et al. [28] and Gupta et al. [19] .

| (m, n) | Frequency (THz) | | |
|----------|-----------------|-----------------|-----------------|
| | Present | Strozzi et al. [28] | Gupta et al. [19] |
| (10, 0) | 8.918 | 8.966 | 8.718 |
| (14, 0) | 6.389 | 6.414 | 6.235 |
| (16, 0) | 5.573 | 5.606 | 5.455 |
| (18, 0) | 4.902 | 4.985 | 4.85 |
| (20, 0) | 4.443 | 4.489 | 4.364 |
| (25, 0) | 3.523 | 3.590 | 3.491 |
| (30, 0) | 2.956 | 2.991 | 2.908 |
| (33, 0) | 2.686 | 2.718 | 2.623 |

rate. Furthermore, the results of the present model are found well validated with published literature.'

**Table 2.4**   Comparison of RRM frequencies of zigzag SWCNTs with Strozzi et al. [28] and Jorio et al. [10] .

| (*m, n*) | Frequency (THz) | | |
|---|---|---|---|
| | Present | Strozzi et al. [28] | Jorio et al. [10] |
| (8, 7) | 6.876 | 6.905 | 7.165 |
| (10, 5) | 6.726 | 6.785 | 7.105 |
| (11, 4) | 6.613 | 6.669 | 6.865 |
| (14, 1) | 6.142 | 6.177 | 6.295 |
| (17, 2) | 4.927 | 4.964 | 5.216 |
| (16, 4) | 4.847 | 4.895 | 5.066 |
| (15, 6) | 4.739 | 4.788 | 4.947 |
| (19, 1) | 4.558 | 4.594 | 4.797 |
| (18, 6) | 4.111 | 4.150 | 4.317 |

## 2.3.1  Vibration effect of length-to-diameter ratios of Armchair SWCNTs

Figures 2.3 and 2.4 show the frequency response of armchair (4, 4) and (11, 11) against the ratio of length-to-diameter, *L/d*. The frequency curves are sketched with different BCs of type C–C and C–SS. Particularly, the natural frequencies corresponding to the ratio of length-to-diameter (*L/d* = 1) for

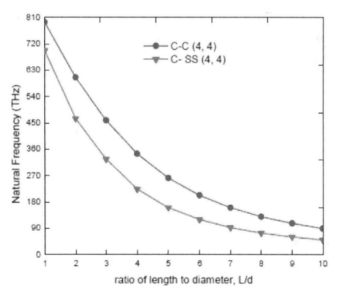

**Figure 2.3**   Ratio of length-to-diameter effect on fundamental frequencies for armchair SWCNTs [C–C, C–SS, (4, 4)].

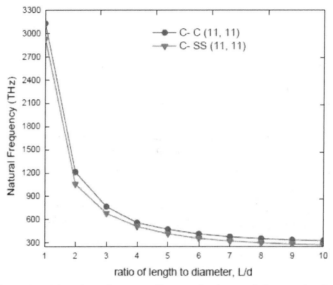

**Figure 2.4** Ratio of length-to-diameter effect on fundamental frequencies for armchair SWCNTs [C–C, C–SS, (11, 11)].

armchair indices (4, 4) and (11, 11) are 799.02 and 3200.34, for the C–C end condition, and 710.31, 2931.37, for the C–SS end condition. Similarly, the frequencies corresponding to the ratio of length-to-diameter ($L/d$ = 2) for armchair index C–C (4, 4) $\sim$ 465.35, C-SS (4, 4) $\sim$ 620.73, and C–C (11, 11) $\sim$ 1203.64, C-SS (11, 11) $\sim$ 1092.48, respectively. It can be seen that the frequencies of C–C, C–SS (11, 11) rapidly decline as compared to the frequencies of C–C, C–SS (4, 4). Now for the frequency pattern for $L/d$ = 10, for armchair indices (4, 4) and (11, 11) are 120.89 and 90.53, for the C–C end condition, and 430.59, 302.91, for the C–SS end condition. In comparison between armchair index (4, 4) and (11, 11), the frequency gap of armchair (4, 4) curves for two boundary conditions is parallel throughout the computation of the ratio of length-to-diameter ($L/d$ = 1–10) while the frequency gap of armchair (11, 11) curves is a different pattern from index (4, 4). The frequencies decrease from ($L/d$ = 1–2) rapidly, while for the length ($L/d$ = 2–4), the frequencies are gently parallel. In the present result, the frequencies meet at $L/d$ (= 5–10). It shows that natural frequencies decrease as $L/d$ is increased, for these boundary conditions. For long SWCNTs, it can be seen that the effect of BCs is insignificant and more prominent at shorter ratio of length-to-diameter ($L/d$ = 1–4) and moderately negligible at ratio of length-to-diameter ($L/d$ = 8–10). Moreover, the frequency pattern of armchair (4, 4) is much interesting due to behavior of frequency look like as same at initial and end of the curves and the gap is larger than its initial and end position. In armchair (4, 4) with prescribed boundary conditions, frequencies are lower than that of armchair (11, 11). Now it is clear that on increasing the armchair indices from (4, 4)–(11, 11), the frequencies increases. In contrast, the frequency curves (11, 11), the mechanism is the same throughout the computations for $L/d$ = 1–10 except $L/d$ = 2–3. The gap is greater for $L/d$ = 2–3 than for other $L/d$. In addition, it is noted that the highest frequency pattern of C-C armchair (4, 4), (11, 11) is greater than C–SS armchair (4, 4), (11, 11). Since a boundary condition is defined by adding some physical constraints, frequency variations are obviously noted for boundary conditions, and they have a very prominent effect on the vibrations of the single-walled carbon nanotubes (SWCNTs). This investigation presents a good application of the wave propagation approach in the arena of CNTs vibrations.

## 2.3.2 Vibration effect of length-to-diameter ratio of zigzag SWCNTs

Figures 2.5 and  2.6 show the frequency response of zigzag (6, 0), (13, 0) against the length-to-diameter ratio with specified boundary conditions. The investigated values of frequencies with two different boundary conditions are at $L/d$ = 1 as C–C = (6, 0) $f$ ~92.873, (13, 0) $f$ ~ 198.567, and for C–SS = (6, 0) $f$ ~90.437, (13, 0) $f$ ~195.821, respectively. It can be seen in the plotted graph the frequency curve decreases suddenly from $L/d$ = 1 to $L/d$ = 2 as C–C = (6, 0) $f$ ~92.873 to C–C = (6, 0) $f$ ~62.453. It is clear that frequency declines from 92.873 to 62.453, which is a long dip. Also, the frequencies for C–C = (13, 0) $f$ ~ 198.567 to C–C = (13, 0) $f$ ~191.864 is a small dip. The same behavior is the same for C–SS boundary condition for both zigzag indices (6, 0), (13, 0). Once again, the frequencies outcomes with proposed boundary conditions at $L/d$ = 2, 3 as C–C = (6, 0) $f$ ~62.453, 58.830, (13, 0) $f$ ~191.864,191.236 and for C–SS = (6, 0) $f$ ~59.213, 56.594, (13, 0) $f$ ~189.584, 189.372 respectively. It can take into account that the zigzag SWCNTs (6, 0), (13, 0), with C–C condition, have the prominent and highest frequencies as compared to C–SS. The frequency pattern with all boundary conditions seems to be parallel for overall values of $L/d$ = 2, 3. As the index of zigzag increases, the frequency must increase. Here as we increase the index from (6, 0) to (13, 0), which is double. Now after careful consideration, the frequencies increase twice from (6, 0) to (13, 0) for both conditions,

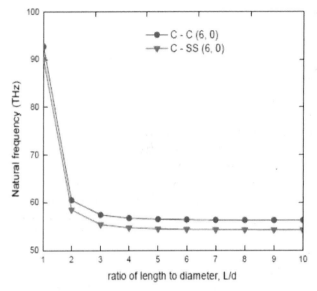

**Figure 2.5**  Ratio of length-to-diameter effect on fundamental frequencies for zigzag SWC-NTs [C–C, C–SS, (6, 0)].

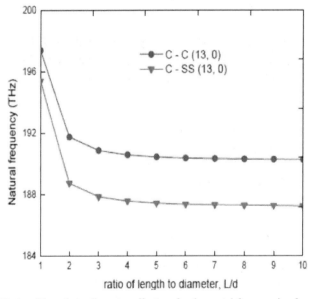

**Figure 2.6**  Ratio of length-to-diameter effect on fundamental frequencies for zigzag SWC-NTs [C–C, C–SS, (13, 0)].

as shown in Figures 2.5 and 2.6. Moreover, the frequency pattern for the length-to-diameter ratio, $L/d$ = 4, 5 is moderately parallel for both conditions. The separation of (C–C, C–SS) frequency curves (6, 0) for $L/d$ = 1 ~ 2 is insignificant, but for (13, 0) is significant. The frequency bandwidth of (6, 0) is small for $L/d$ = 2–10, and this bandwidth of (13, 0) is four times larger. Now there is an interesting phenomenon: minor frequency change from $L/d$ = 3–4. After that, there is no change in the frequencies from $L/d$ = 4–10 for varying index (6, 0)–(13, 0). Constant behavior is observed from $L/d$ = 4 ~ 10, and frequencies are linear with two conditions. It is exhibited that. Finally, Figures 2.5 and 2.6 shows that natural frequencies decrease as the ratio of length-to-diameter increase. For small zigzag single-walled carbon nanotubes, it can be seen that the effect of frequencies is less significant and more prominent for a long tube. Also, the boundary condition has a momentous effect on the vibration of single-walled carbon nanotubes. These boundary conditions are kept on the edge of the tube. It is noted that the frequency curves of C–SS boundary condition are the lowest outcomes as compared to C–C boundary condition. This is due to the physical constraints of boundary conditions.

### 2.3.3 Vibration effect of length-to-diameter ratios of chiral SWCNTs

The natural frequencies computed in Figures 2.7 and 2.8 for chiral single-walled carbon nanotubes against the ratio of length-to-diameter with two different boundary conditions, namely, clamped-clamped (C-C) and clamped-simply supported (C-SS). The chiral indices are used here by changing the values of $m$ and $n$. For the first frequency of C–C (7, 4), C–SS (7, 4), $f$(THz) is 1200.387, 1199.210, and also for C–C (10, 6), C–SS (10, 6), $f$(THz) is 2998.220, 1678.821 which tends to decrease very fastly and gradually, respectively. For the second and third frequency C–C (7, 4), C–SS (7, 4), $f$(THz) are 786.412–651.023, 658.910–573.291, the frequencies reduces and bends to another shape, and also for C–C (10, 6), C–SS (10, 6), $f$(THz) are 1499.228–852.473, 1023.129–587.431 which tends to decrease as it decreases for first frequency. Also, for the third to the fourth frequency of C–C (10, 6), C–SS (10, 6), the behavior of $f$(THz) bends suddenly. After the third frequency in the case of chiral (7, 4), and fourth frequency of chiral (10, 6), the frequencies are smooth and parallel till $L/d$ = 10. It can also be seen that the distance between upper and lower curve is small for the first two frequencies and remain the same till $L/d$ = 10 for chiral (7, 6). Now for

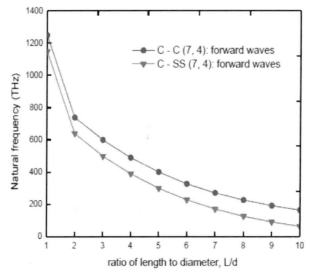

**Figure 2.7** Ratio of length-to-diameter effect on fundamental frequencies for chiral SWC-NTs [C-C, C-SS, (7, 4)].

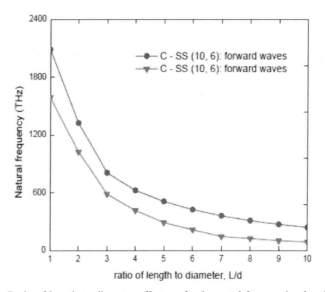

**Figure 2.8** Ratio of length-to-diameter effect on fundamental frequencies for chiral SWC-NTs [C–C, C–SS, (10, 6)].

chiral (10, 6), the sandwich part of upper and lower curves is uniform for the first three frequencies, and from the fourth frequency, gap increases till the eighth frequency. It is too interesting that the frequency curve distance is small from the ninth and tenth curves. It is due to the vibrational behavior of single-walled carbon nanotubes placed on different edges. From Figures 2.7 and 2.8, it is observed as on increasing the indices of the tube (7, 4)–(10, 6), the frequencies tend to increase.

Comparing Figures 2.3– 2.8 shows that frequencies are not affected by changing the structure of single-walled carbon nanotubes as armchair, zigzag, and chiral. The frequencies increase by decreasing the ratio of length-to-diameter ratio. Also, the frequencies increase when the index increases as armchair indices (4, 4)– (11, 11), for varying index (6, 0)– (13, 0), and indices of the chiral tube (7, 4)– (10, 6). This frequency behavior remains the same for increasing the length-to-diameter ratio throughout the computation in this chapter.

## 2.4 Conclusion

In this paper, Flügge's shell theory based on Rayleigh–Ritz method is developed for the vibration of single-walled carbon nanotubes. The frequencies of three sorts of SWCNTs are conducted with two boundary conditions. The resulting frequencies are gained for length-to-diameter ratios. The natural frequency becomes more prominent for lower length-to-diameter ratios and diminished for higher ratios. Different indices of armchair (4, 4), (11, 11), zigzag (6, 0), (13, 0), and chiral (7, 4), (10, 6) are investigated. The frequencies increase by decreasing the ratio of length-to-diameter. Also, the frequencies increase when the index increases as armchair indices (4, 4)–(11, 11), for varying index (6, 0)–(13, 0), and indices of the chiral tube (7, 4)–(10, 6). This frequency behavior remains the same for increasing the length-to-diameter ratio throughout the computation in this chapter. It is observed that frequencies are not affected by changing the structure of single-walled carbon nanotubes as armchair, zigzag, and chiral. Also, the boundary condition has a momentous effect on the vibration of single-walled carbon nanotubes. These boundary conditions are kept on the edge of the tube. It is noted that the frequency curves of C–SS boundary condition are the lowest outcomes as compared to C–C boundary condition. This is due to the physical constraints of boundary conditions.

# References

[1] S. Iijima, "Helical microtubules of graphitic carbon," *nature*, vol. 354, no. 6348, pp. 56–58, 1991.

[2] Ebbesen, T. W., & Ajayan, P. M. (1992). Large-scale synthesis of carbon nanotubes. *Nature, 358*(6383), 220–222.

[3] Iijima, S., & Ichihashi, T. (1993). Single-shell carbon nanotubes of 1-nm diameter. *nature, 363*(6430), 603–605.

[4] Dujardin, E., Ebbesen, T. W., Hiura, H., & Tanigaki, K. (1994). Capillarity and wetting of carbon nanotubes. *Science, 265*(5180), 1850-1852.

[5] Ruoff, R. S., & Lorents, D. C. (1995). Mechanical and thermal properties of carbon nanotubes. *carbon, 33*(7), 925=-930.

[6] Treacy, M. J., Ebbesen, T. W., & Gibson, J. M. (1996). Exceptionally high Young's modulus observed for individual carbon nanotubes. Nature, 381(6584), 678–680

[7] Peigl, L. W. T., (1997). The NURBS book. Netherland: Springer.

[8] Lordi, V. & Yao. N. (1998). Young's modulus of single-walled carbon nanotubes. J. Appl. Phys., 84, 1939–1943.

[9] Krishnan, A., Dujardin, E., Ebbesen, T. W., Yianilos, P. N., & Treacy. M. M. J. (1998). Young's modulus of single-walled nanotubes. Physical Review B (Condensed Matter and Materials Physics), 58(20), 14013–14019.

[10] Jorio, A., Saito, R., Hafner, J. H., Lieber, C. M., Hunter, D. M., McClure, T., ... & Dresselhaus, M. S. (2001). Structural (n, m) determination of isolated single-wall carbon nanotubes by resonant Raman scattering. *Physical Review Letters, 86*(6), 1118.

[11] Zhang, X. M., Liu, G. R., & Lam, K. Y. (2001). Coupled vibration analysis of fluid-filled cylindrical shells using the wave propagation approach, Applied Acoustics, 62, 229–243.

[12] Li, C., & Chou, T. W. (2003). A structural mechanics approach for the analysis of carbon nanotubes. International Journal of Solids and Structures, 40(10), 2487–2499.

[13] Pantano, A., Parks, D. M., & Boyce, M. C. (2004). Mechanics of deformation of single-and multi-wall carbon nanotubes. Journal of the Mechanics and Physics of Solids, 52(4), 789–821.

[14] Wang, Q., & Varadan, V. K. (2007). Application of nonlocal elastic shell theory in wave propagation analysis of carbon nanotubes. Smart Materials and Structures, 16(1), 178.

[15] Wang, C. Y. & Zhang, L. C. (2007). Modeling the free vibration of single-walled carbon nanotubes. 5th Australasian Congress on Applied Mechanics, ACAM, Brisbane, Australia, 10–12.

[16] Natsuki, T., Ni, Q. Q., & Endo, M. (2008). Analysis of the vibration characteristics of double-walled carbon nanotubes. Carbon, 46(12), 1570-1573.

[17] Elishakoff, I., & Pentaras. D. (2009). Fundamental natural frequencies of double-walled carbon nanotubes. Journal of Sound and Vibration, 322, 652–664.

[18] Lee, H. L., & Chang, W. J. (2009). Vibration analysis of fluid-conveying double-walled carbon nanotubes based on nonlocal elastic theory. Journal of Physics: Condensed Matter, 21(11), 115302.

[19] Gupta, S. S., Bosco, F. G., & Batra, R. C. (2010). Wall thickness and elastic moduli of single-walled carbon nanotubes from frequencies of axial, torsional and inextensional modes of vibration. *Computational Materials Science*, *47*(4), 1049-1059.

[20] Simsek, M. (2010). Vibration analysis of a single-walled carbon nanotube under action of a moving harmonic load based on nonlocal elasticity theory. Physica E, 43,182–191.

[21] Yang, J., Ke, L. L., & Kitipornchai, S. (2010). Nonlinear free vibration of single-walled carbon nanotubes using nonlocal Timoshenko beam theory. Physica E: Low-dimensional Systems and Nanostructures, 42(5), 1727–1735.

[22] Thai, H.T. (2012). A nonlocal beam theory for bending, buckling, and vibration of nanobeams. Int. J. Eng. Sci., 52, 56–64.

[23] Rafiee, M., Yang, J., & Kitipornchai. S. (2013). Large amplitude vibration of carbon nanotube reinforced functionally graded composite beams with piezoelectric layers. Composite Structures, 96, 716–725.

[24] Ansari, R., & Arash, B. (2013). Nonlocal Flügge shell model for vibrations of double-walled carbon nanotubes with different boundary conditions. Journal of Applied Mechanics, 80(2), 021006.

[25] Chawis, T., Somchai, C., & Li, T. (2013). Nonlocal theory for free vibration of single-walled carbon nanotubes, Advan ced Materials Research, 747, 257–260.

[26] Ansari, R., Rouhi, S., & Aryayi, M. (2013). Nanoscale finite element models for vibrations of single-walled carbon nanotubes: atomistic versus continuum. Applied Mathematics and Mechanics, 34(10), 1187-1200.

[27] Ponnusamy, P., & Amuthalakshmi, A. (2013, July). Vibration analysis of a viscous fluid conveying double walled carbon nanotube. In International Conference on Advanced Nanomaterials & Emerging Engineering Technologies (pp. 231-236). IEEE.

[28] Strozzi, M., Manevitch, L. I., Pellicano, F., Smirnov, V. V., & Shepelev, D. S. (2014). Low-frequency linear vibrations of single-walled carbon nanotubes: Analytical and numerical models. *Journal of Sound and Vibration*, *333*(13), 2936-2957.

[29] Kiani, K. (2014). Vibration and instability of a single-walled carbon nanotube in a three dimensional magnetic field. Journal of Physics and Chemistry of Solids, 75(1), 15–22.

[30] Rouhi. H., Bazdid-Vahdati, M, & Ansari, R. (2015). Rayleigh-Ritz vibrational analysis of multi-walled carbon nanotubes based on the non-local Flügge shell theory. Journal of Composites, 10.1155/2015/750392.

[31] Rakrak, K., Zidour, M., Heireche, H., Bousahla, A. A., & Chemi, A. (2016). Free vibration analysis of chiral double-walled carbon nanotube using non-local elasticity theory. Advances in Nano Research, 4(1), 031.

[32] Ehyaei, J., & Daman, M. (2017). Free vibration analysis of double walled carbon nanotubes embedded in an elastic medium with initial imperfection. Advances in Nano Research, 5(2), 179.

[33] Kumar, B. R. (2018). Investigation on mechanical vibration of double-walled carbon nanotubes with inter-tube Van der waals forces. *Advances in nano research*, 6(2), 135.

[34] Sobamowo, M. G., Akanmu, J. O., Adeleye, O. A., & Yinusa, A. A. (2019). Nonlinear vibrations of single-and double-walled carbon nanotubes resting on two-parameter foundation in a magneto-thermal environment. *SN Applied Sciences*, *1*(10), 1-22.

[35] Miyashiro, D., Taira, H., Hamano, R., Reserva, R. L., & Umemura, K. (2020). Mechanical vibration of single-walled carbon nanotubes at different lengths and carbon nanobelts by modal analysis method. *Composites Part C: Open Access*, 2, 100028.

[36] Jena, S. K., Chakraverty, S., Malikan, M., & Tornabene, F. (2021). Stability analysis of single-walled carbon nanotubes embedded in winkler foundation placed in a thermal environment considering the surface effect using a new refined beam theory. *Mechanics Based Design of Structures and Machines*, 49(4), 581-595.

[37] Abdullah, S. S., Hosseini-Hashemi, S., Hussein, N. A., & Nazem-nezhad, R. (2022). Effect of temperature on vibration of cracked single-walled carbon nanotubes embedded in an elastic medium under

different boundary conditions. *Mechanics Based Design of Structures and Machines*, *50*(5), 1614-1639.

[38] Shi, X.; Pugno, N.M.; Cheng, Y.; Gao, H. Gigahertz breathing oscillators based on carbon nanoscrolls. *Appl. Phys. Lett.* 2009, *95*, 163113.

[39] Amabili, M. (2008). *Nonlinear vibrations and stability of shells and plates*. Cambridge University Press.

[40] Selim, M. M. (2010). Torsional vibration of carbon nanotubes under initial compression stress. *Brazilian Journal of Physics*, *40*, 283-287.

[41] FORSBERG, K., "Influence of Boundary Conditions on the Model Characteristics of Thin Cylindrical Shells", AIAA Journal, Vol. 2, No. 12, 1964, p. 2150-2157.

[42] KRAUS, H., "Thin Elastic Shells", John Wiley and Sons, New York, 1967.

[43] Ansari, R., & Rouhi, H. (2015). Nonlocal Flügge shell model for the axial buckling of single-walled Carbon nanotubes: An analytical approach. *International Journal of Nano Dimension*, *6*(5), 453-462.

[44] Zhang, Y. Y., Wang, C. M., & Tan, V. B. C. (2009). Assessment of Timoshenko beam models for vibrational behavior of single-walled carbon nanotubes using molecular dynamics. *Advances in Applied Mathematics and Mechanics*, *1*(1), 89-106.

# 3

# Accuracy of Stiffness on the Vibration of Single-walled Carbon Nanotubes: Orthotropic Shell Model

## Abstract

This chapter represents study of the vibration characteristics of single-walled carbon nanotubes (SWCNTs) for two boundary conditions. The wave propagation approach is used as a numerical method to solve the governing equations of the orthotropic shell model. Simply supported and clamped-free boundary conditions are utilized to investigate the natural frequencies. The computer software MATLAB is used for the extraction of results. Based on this model, the frequency effect of stiffness with boundary conditions is discussed and examined. The frequency of armchair, zigzag, and chiral zigzag single-walled carbon nanotubes (SWCNTs) is observed against the stiffness with two prescribed boundary conditions. The frequencies increase on increasing the stiffness of the carbon nanotubes. The frequency jumps vertically upward. These sudden jumps show that the material of the tube is very stiff during vibration. It can be seen that the frequencies of clamped-free boundary condition is lower than simply supported condition. The present results are validated with the exact results and experimental results of Raman Spectroscopy in tabular form. The comparison shows that the vibration responses of SWCNTs are influenced by frequency variation with stiffness. This study should be helpful for the scientists working on small size of devices in the industry.

**Keywords:** Orthotropic shell model, SWCNTs, MATLAB, wave propagation approach, Raman spectroscopy, stiffness.

## 3.1 Introduction

Carbon was discovered in 1779 in the graphitic form. The discovery of carbon nanotubes as a novel production is a significant growth reported by Iijima [1]. This structure has a carbon sheet containing more than two concentric circles. Iijima [1] and his coworkers discovered multi-walled carbon nanotubes which have several nested cylinders with interlayer spacing. Iijima et al. [2] presented evidence for the growth of open-end carbon nanotubes producing different morphologies involving many pairs. Bethune et al. [3] independently found an efficient way of synthesizing single-walled carbon nanotubes (SWCNTs). The following decades saw a surge in the volume of research dedicated to this field of nanotechnology, studying their exceptional properties arising due to its high length-to-diameter ratio and exploring their applications in a variety of fields, including but not limited to, micro-electronics, energy, mechanical, and biological applications [4, 5]. Iijima et al. [6] performed successful high-resolution electron microscopic investigation and simulation on the structural flexibility of single and multi-walled CNTs under mechanical pressure. Kane and Mele [7] investigated the consequence of shape variation with respect to electronic possessions. In the last decade, most research work regarding vibrations of double-walled CNTs has been produced using various beam theories. Nanotubes, like cylindrical molecular systems at nano/micro scale, possess wavy configuration, and a powerful mathematical model is needed to study such systems. Krishnan et al. [8] assessed the Young moduli of single-walled CNTs by noticing their autonomous room-temperature vibration and response through a transmission electron microscope (TEM). Electron nano graphs have been used to measure vibration amplitudes and tube dimensions. Nardeli et al. [9] developed the ideal form of CNTs through the hexagonal ring and $sp_2$ hybrid carbon particles. Stone-Wales defects composed of five-membered rings are an example of hexagone-heptagone.

Mickelson et al. [10] restored the original structure of SWCNTs after integrating the polymer matrix. For the application of CNTs, functional groups can improve compatibility with inorganic compounds. Miyamoto [11] practiced the density functional theory and showed the modifications in the electronic state near to fermi level due to inter wall interactions of double-walled CNTs. Their calculations also pointed out the shifting of the energy band to 0.3 eV depending on the diameter of each tube. Harik [12] studied the behaviors of CNTs using beam models. He concluded that the beam model could be used for the qualitative exploration of CNTs because the nanotube aspect ratio is greater than 10 nm. The behavior of elements becomes more complicated using experimental, molecular dynamics, and continuum models. Flahaut et al. [13] introduced the first-time-ever synthesis of double-walled CNTs based on gram-scale using the chemical vapor decomposition (CVD) approach. Li and Chou [14] produced SWCNTs based-sensors with the principal of sensing for measuring the strain and pressure. By using atomistic modeling and molecular structural mechanics methods, the carbon nanotube in a bridged configuration is stimulated. Yoon et al. [15] utilized frequencies that fall in the tera-hertz range using the Timoshenko beam model and also used the EBM to find the aspect ratio of DWCNTs. In particular, they compared the significant effect of the Euler and Timoshenko beam models on small-diameter and large-diameter DWCNTs. They suggested for the vibration of short CNTS, the TBM for tera-hertz is relevant rather than EBM. Wang and Varadan [16] established a nonlocal continuum model based on elastic beam theory to observe the small-scale effect on vibration of both single and double-walled CNTs. He et al. [17] investigated the vibration frequencies of MWCNTs based on Donnell shell theory, and these explicit formulas obtain the radial frequencies of CNTs. The results showed that frequencies are small and can be ignored due to the van der Waals effect. The domination of rotary inertia was described by using Timoshenko beam theory through the vibrational behavior of multi-walled CNTs [18]. Governing equations associated with boundary conditions to realize ratios were designed by DQM. The cross-check of results by EBM determined the fact that frequency tends to be overestimated with a miniscule aspect ratio. Consequently, the Timoshenko beam model for suchlike problems and phenomena was a better choice for predicting nanotubes' frequencies. Duan et al. [19] presented the calibrated values of small-scale parameters via Timoshenko beam theory with MD simulation to produce the vibrating frequency of single-walled CNTs. The calibration of nonlocal parameters ranges from 0 to 19 rather than to be specified/particular, as reported earlier. The calibrated range of nonlocal parameters depends upon different mode shapes, end supports, and length-to-diameter ratio. It is considered more useful in studying vibrational aspects of nano/ micro rods/ tubes/ beams by the suggested nonlocal beam model. Natuski et al. [20] adopted a wave propagation approach to investigate single- and double-walled CNTs brimming with fluids. Flügge shell theory was proposed to form governing equations of motion for CNTs. They observed that the free path for phonons is an important factor that decides thermal conductivity. Wang et al. [21] used the beam and shell model for flexural waves in SWCNTs and MWCNTs based on wave propagation. Zhang et al. [22] studied the assessment of TBM for the flexural vibration of SWCNTs using reactive empirical bond order. Vibration frequencies are assessed using MD simulation, and the results show that chirality is much affected for higher frequencies. The effects of aspect ratios against fundamental frequencies are determined using TBM with initial strain, chiral angles, and boundary conditions.

Elishkoff and Pentaras [23] applied the Petrov-Galerkin and Bubnov-Galerkin methods to simulate the frequencies of CNTs and argued the explicit formula with different edge conditions. Khademolhosseini et al. [24] studied torsional buckling using nonlocal elasticity shell models. Madihav et al. [25] developed the model utilized at the same time as Timoshenko and Euler Bernoulli beam theories with the help of harmonic balanced approach. They introduced the two-way vibrational behavior along with Timoshenko beam model of single-walled CNTs. Also, they observed the frequency spectra with adjacent and medium tubes of double-walled CNTs under the effects of van der Waal forces. Khademolhosseini et al. [26] also studied the size effects in the dynamic torsional response of a single-walled carbon nanotube by modeling it as a modified nonlocal continuum shell.

Basirjafri et al. [27] obtained the radial breathing mode (RBM) frequency by using thin shell theory in reliance on Hamilton's principle of single-walled CNTs, and the results included the influence on the variation of Poisson's ratio. The study considered the influence of distinct boundary conditions and tube chirality. They concluded that an increase in tube length, diameter, and change in the atomic scheme was responsible for variation in obtained frequencies. Mikhasev [28] studied the vibrations of single-walled carbon nanotubes embedded in a nonhomogeneous elastic matrix using nonlocal continuum shell model. Chemi et al. [29] scrutinized the effect of buckling under axial compression of chiral double-walled CNTs. Nonlocal Timoshenko beam model has been formulated by incorporating the nonlocal elasticity equations. The study observed the length-to-diameter ratio and increase in value of scale coefficient as major influential factors to critical buckling loads of tubes. It also indicated that critical buckling loads become more sensitive for a higher value of nonlocal parameter and shortened tubes. Wang and Gao [30] investigated wave propagation in microtubules based on an orthotropic shell model. The effect of small-scale with longitudinal wave vector is studied using nonlocal elastic theory. It is seen that the influence of small-scale is significant for smaller wave length. Soltani [31] investigated the nonlinear vibrational characteristics of SWCNTs using the theory of nonlocal elasticity and Karman's geometric nonlinearity theory. The controlling equation is derived from DST, and partial differential equations (PDEs) are converted into differential equations by invoking Galerkin's technique (GT). The influence of aspect ratios, nonlocal, nonlinear, and circumferential parameters are investigated. Bouadi et al. [32]developed a new model displacement field for the nonlocal buckling properties of single graphene sheet. The Eringen relation was used for the theoretical formation with the length scale parameter.

Fatahi-Vajari et al. [33] analyzed the torsional vibration of SWCNTs based on second-order PDEs. These equations are reduced by GT and the torsional frequency equation is obtained by using HPM. Malikan and Eremeyev [34] predicted the buckling analysis of CNTs using the Winkler matrix with different boundary conditions. Hamilton's principle was used to measure the hardness- and softness-stiffness of CNTs. Pourasghar et al. [35] presented the transient heat conduction and vibration of SWCNTs based on Eringen's and heat conduction theories. The scale of energy is defined using the nonlocal term with heat conduction theory. Ghasemi and Gouran [36] assessed the vibration of SWCNTs filled with fluid resting on the Pasternak foundation. The DTM is utilized to solve differential equations.

The present study uses the orthotropic shell model to investigate the vibration response of armchair, zigzag, and chiral SWCNTs. The wave propagation approach (WPA) is developed to generate eigenvalue form with the help of the axial modal function in matrix representation. The formulation of WPA is given by Zhang et al. [37] a brief yet simple. There is a huge importance of vibration analysis in its practical nature. In engineering, the vibration effect is very stimulation for structural material. Vibration is done in that body that has some elasticity and mass. In terms of elasticity, the stiffness occurs. When this stiffness exceeds a limit, then the structure can occur as a failure. To escape from such failure system, this study is developed to check the vibration of that system in increasing or decreasing the effect of stiff with frequencies, and this stiff is overcome by applying the boundary conditions, which is our particular motive. Here only two boundary conditions viz simply supported and clamped-free boundaries are used for maintaining the stiff of armchair, zigzag, and chiral SWCNTs. The framework of this model is to check the frequency effect of increasing the stiffness of the material.

## 3.2 Model-based Method

### 3.2.1 Orthotropic shell model

In present model, the orthotropic shell model is presented for the vibration of single walled carbon nanotubes. Fig. 3.1 shows the coordinate system of

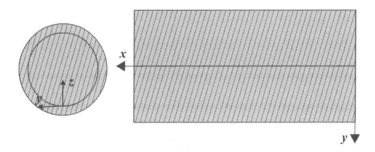

**Figure 3.1**   Coordinate system of shell.

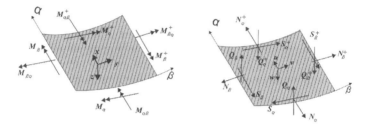

**Figure 3.2**   Force analysis of shell element.

the tube which look like as a cylindrical shell (See Fig. 2.1, chapter 2) and Figure 3.2 presented the force analysis of shell element [30]. The cylindrical coordinates system are $(x, y, z)$, where $(x, y)$ are coordinates acts as axial, circumferential and radial coordinate is $z$.

The dimensionless coordinates are $(\alpha, \beta, \gamma)$ which can be represented as $(\alpha = x/R, \beta = y/R, \gamma = z/R)$. The displacement field of middle surface are $(u, v, w)$ in the directions of $(\alpha, \beta, \gamma)$.

## 3.2.2 Relationships between strain and displacement

In this theory, the geometrical relationships of strain and displacement are based on Ref [38, 39]. The dimensionless strains $(\varepsilon_\alpha, \varepsilon_\beta, \varepsilon_{\alpha\beta})$ of middle surface is taken into account as

$$\varepsilon_\alpha = \frac{1}{R} \left( \frac{\partial u}{\partial \alpha} - \gamma \frac{\partial^2 w}{\partial \alpha^2} \right) \tag{3.1}$$

$$\varepsilon_\beta = \frac{1}{R}\left(\frac{\partial v}{\partial \beta} + w\right) - \frac{\gamma}{R(1+\gamma)}\left(\frac{\partial^2 w}{\partial \beta} + w\right) = 0 \qquad (3.2)$$

$$\varepsilon_{\alpha\beta} = \frac{\gamma}{R(1+\gamma)}\left[\frac{\partial u}{\partial \beta} + \frac{\partial v}{\partial \alpha} + 2\gamma\left(\frac{\partial v}{\partial \alpha} - \frac{\partial^2 w}{\partial \alpha \partial \beta}\right) + \gamma^2\left(\frac{\partial v}{\partial \alpha} - \frac{\partial^2 w}{\partial \alpha \partial \beta}\right)\right] = 0 \qquad (3.3)$$

### 3.2.3 Relationships between stress and strain

According to plane stress hypothesis based on Hook's law, the dimensionless stresses $(\sigma_\alpha, \sigma_\beta, \tau_{\alpha\beta})$ related to the dimensionless strains $(\varepsilon_\alpha, \varepsilon_\beta, \varepsilon_{\alpha\beta})$ of middle surface are derived as [40].

$$\sigma_\alpha = E_1\frac{(\varepsilon_\alpha + \mu_1\varepsilon_\beta)}{(1 - \mu_1\mu_2)} \qquad (3.4)$$

$$\sigma_\beta = E_2\frac{(\mu_2\varepsilon_\alpha + \varepsilon_\beta)}{(1 - \mu_1\mu_2)} \qquad (3.5)$$

$$\tau_{\alpha\beta} = G\varepsilon_{\alpha\beta} \qquad (3.6)$$

Where $\sigma_\alpha\,(\varepsilon_\alpha)$, $\sigma_\beta\,(\varepsilon_\beta)$ are normal stress (strain) and $\tau_{\alpha\beta}\,(\varepsilon_{\alpha\beta})$ are shear stresses (strains). The elastic moduli and Poisson's ratios are represented by $E_1, E_2$ and $\mu_1, \mu_2$, respectively, in the directions of $\alpha, \beta$. The shear modulus is denoted by $G$ and the combination of elastic moduli and Poisson's ratios is $E_1\mu_1 = E_2\mu_2$. The Laplace operator $(\nabla^2)$ in the dimensionless coordinates form is designated as $(\partial^2/\partial\alpha^2 + \partial^2/\partial\beta^2)/R^2$.

The shell elements in the dimensionless coordinate system, the stress resultants $(N, S, Q)$ and moment $M$. The pre-stress condition oriented from thermal expansion is ignored in the presence of temperature and taken as reference temperature. Thus the dynamically equilibrium equations are as under.

$$\frac{\partial N_\alpha}{\partial \alpha} + \frac{\partial S_\beta}{\partial \beta} + k = \rho h R\frac{\partial^2 u}{\partial t^2} \qquad (3.7)$$

$$\frac{\partial N_\beta}{\partial \beta} + \frac{\partial S_\alpha}{\partial \alpha} + Q_\beta = \rho h R\frac{\partial^2 v}{\partial t^2} \qquad (3.8)$$

$$\frac{\partial Q_\alpha}{\partial \alpha} + \frac{\partial Q_\beta}{\partial \beta} - N_\beta = \rho h R\frac{\partial^2 w}{\partial t^2} \qquad (3.9)$$

$$\frac{\partial M_{\alpha\beta}}{\partial \alpha} + \frac{\partial M_{\beta}}{\partial \beta} - RQ_{\beta} = 0 \tag{3.10}$$

$$\frac{\partial M_{\beta\alpha}}{\partial \beta} + \frac{\partial M_{\alpha}}{\partial \alpha} - RQ_{\alpha} = 0 \tag{3.11}$$

Where $\rho$ is mass density.

The resultants derived from stress components (Eqs. 3.4–3.6) are followed as

$$\begin{bmatrix} N_{\alpha} & S_{\alpha} \\ M_{\alpha} & M_{\alpha\beta} \end{bmatrix} = \int_{-h/2}^{h/2} \begin{bmatrix} \sigma_{\alpha} & \tau_{\alpha\beta} \\ z\sigma_{\alpha} & z\tau_{\alpha\beta} \end{bmatrix} \left(1 + \frac{z}{R}\right) dz \tag{3.12}$$

$$\begin{bmatrix} N_{\beta} & S_{\beta} \\ M_{\beta} & M_{\beta a} \end{bmatrix} = \int_{-h/2}^{h/2} \begin{bmatrix} \sigma_{\beta} & \tau_{\beta\alpha} \\ z\sigma_{\beta} & z\tau_{\beta\alpha} \end{bmatrix} dz \tag{3.13}$$

$$[Q_{\alpha}, Q_{\beta}] = \int_{-h/2}^{h/2} [\tau_{\alpha z}, \tau_{\beta z}] \, dz \tag{3.14}$$

The shell thickness is $h$, and these equations can be written as

$$N_{\alpha} = \frac{K}{R} \left[ \frac{\partial u}{\partial \alpha} + \mu_1 \left( \frac{\partial v}{\partial \beta} + w \right) - c^2 \left( \frac{\partial^2 w}{\partial^2 \alpha} \right) \right] \tag{3.15}$$

$$N_{\beta} = \frac{Kk_1}{R} \left[ \frac{\partial v}{\partial \beta} + \mu_2 \frac{\partial u}{\partial \alpha} + w + c^2 \left( \frac{\partial^2 w}{\partial \beta^2} + w \right) \right] \tag{3.16}$$

$$S_{\alpha} = \frac{Kk_2}{R} \left[ \frac{\partial u}{\partial \beta} + \frac{\partial v}{\partial \alpha} - c^2 \left( \frac{\partial^2 w}{\partial \alpha \partial \beta} - \frac{\partial v}{\partial \alpha} \right) \right] \tag{3.17}$$

$$S_{\beta} = \frac{Kk_2}{R} \left[ \frac{\partial u}{\partial \beta} + \frac{\partial v}{\partial \alpha} + c^2 \left( \frac{\partial^2 w}{\partial \alpha \partial \beta} + \frac{\partial v}{\partial \alpha} \right) \right] \tag{3.18}$$

$$M_{\alpha} = -Kc^2 \left[ \frac{\partial u}{\partial \alpha} + \mu_1 \frac{\partial v}{\partial \beta} - \left( \frac{\partial^2 w}{\partial \alpha^2} + \mu_1 \frac{\partial^2 w}{\partial \beta^2} \right) \right] \tag{3.19}$$

$$M_{\beta} = Kk_1 c^2 \left[ \frac{\partial^2 w}{\partial \beta^2} + w + \mu_2 \frac{\partial^2 w}{\partial \alpha^2} \right] \tag{3.20}$$

$$M_{\alpha\beta} = 2Kk_2 c^2 \left[ \frac{\partial v}{\partial \alpha} - \frac{\partial^2 w}{\partial \alpha \partial \beta} \right] \tag{3.21}$$

$$M_{\beta a} = Kk_2 c^2 \left[ \frac{\partial u}{\partial \beta} - \frac{\partial v}{\partial \alpha} + 2 \frac{\partial^2 w}{\partial \alpha \partial \beta} \right] \tag{3.22}$$

$$Q_\alpha - \frac{Kc^2}{R}\left[\frac{\partial^2 u}{\partial \alpha^2} - k_2 \frac{\partial^2 u}{\partial \beta^2} + (k_2 + \mu_1)\frac{\partial^2 v}{\partial \alpha \partial \beta} - \frac{\partial^3 w}{\partial \alpha^3} - (2k_2 + \mu_1)\frac{\partial^3 w}{\partial \alpha \partial \beta^2}\right]$$
(3.23)

$$Q_\beta = \frac{Kk_1 c^2}{R}\left[2\frac{k_2}{k_1}\frac{\partial^2 v}{\partial \alpha^2} - \frac{\partial^3 w}{\partial \beta^3} - \frac{\partial w}{\partial \beta} - \left(2\frac{\mu_1}{k_1} + \mu_2\right)\frac{\partial^3 w}{\partial \alpha^2 \partial \beta}\right]$$
(3.24)

where

$$K = \frac{E_1 h}{(1 - \mu_1 \mu_2)},\ k_1 = \frac{E_1}{E_2},\ k_2 = \frac{G(1 - \mu_1 \mu_2)}{E_1},\ c^2 = \frac{h_0^3}{12R^2 h}$$

The effective thickness of shell is denoted as $h_0$.

By using the resultants of eqn (3.12)–(3.24) in eqn (3.7)–(3.11), the set of three PDE is attained.

$$\left[\frac{\partial^2}{\partial \alpha^2} + k_2\left(1 + c^2\right)\frac{\partial^2}{\partial \beta^2}\right]u + \left[(\mu_1 + k_2)\frac{\partial^2}{\partial \alpha \partial \beta}\right]v$$
$$+ \left[\mu_1 \frac{\partial}{\partial \alpha} + c^2\left(k_2\frac{\partial^3}{\partial \alpha \partial \beta^2} - \frac{\partial^3}{\partial \alpha^3}\right)\right]w = \frac{\rho h R^2}{K}\frac{\partial^2 u}{\partial t^2}$$
(3.25)

$$\left[(\mu_1 + k_2)\frac{\partial^2}{\partial \alpha \partial \beta}\right]u + \left[k_2\left(1 + 3c^2\right)\frac{\partial^2}{\partial \alpha^2} + k_1\frac{\partial^2}{\partial \beta^2}\right]v+$$
$$\left[k_1 \frac{\partial}{\partial \beta} - c^2\left(\mu_1 + 3k_2\right)\frac{\partial^3}{\partial \alpha^2 \partial \beta}\right]w = \frac{\rho h R^2}{K}\frac{\partial^2 v}{\partial t^2}$$
(3.26)

$$\left[\mu_1\frac{\partial}{\partial \alpha} - c^2\left(\frac{\partial^3}{\partial \alpha^3} - k_2\frac{\partial^3}{\partial \alpha \partial \beta^2}\right)\right]u + \left[k_1\frac{\partial}{\partial \beta} - c^2\left(\mu_1 + 3k_2\right)\frac{\partial^3}{\partial \alpha^2 \partial \beta}\right]v$$
$$+ \left[\left(1 + \frac{1}{c^2}\right)k_1 + \frac{\partial^4}{\partial \alpha^4} + k_1\frac{\partial^4}{\partial \beta^4} + 2k_1\frac{\partial^2}{\partial \beta^2} + (2\mu_1 + 4k_2)\frac{\partial^4}{\partial \alpha^2 \partial \beta^2}\right]$$
$$w = \frac{\rho h R^2}{K}\frac{\partial^2 w}{\partial t^2}$$
(3.27)

The new form of eqn (3.25)–(3.27) is

$$\left[\frac{\partial^2 u}{\partial \alpha^2} + k_2\left(1 + c^2\right)\frac{\partial^2 u}{\partial \beta^2}\right] + \left[(\mu_1 + k_2)\frac{\partial^2 v}{\partial \alpha \partial \beta}\right]$$
$$+ \left[\mu_1 \frac{\partial w}{\partial \alpha} + c^2\left(k_2\frac{\partial^3 w}{\partial \alpha \partial \beta^2} - \frac{\partial^3 w}{\partial \alpha^3}\right)\right] = \frac{\rho h R^2}{K}\frac{\partial^2 u}{\partial t^2}$$
(3.28)

$$\left[ (\mu_1 + k_2) \frac{\partial^2 u}{\partial\alpha\partial\beta} \right] + \left[ k_2 \left( 1 + 3c^2 \right) \frac{\partial^2 v}{\partial\alpha^2} + k_1 \frac{\partial^2 v}{\partial\beta^2} \right]$$
$$+ \left[ k_1 \frac{\partial w}{\partial\beta} - c^2 \left( \mu_1 + 3k_2 \right) \frac{\partial^3 w}{\partial\alpha^2\partial\beta} \right] = \frac{\rho h R^2}{K} \frac{\partial^2 v}{\partial t^2}$$

(3.29)

$$\left[ \mu_1 \frac{\partial u}{\partial\alpha} - c^2 \left( \frac{\partial^3 u}{\partial\alpha^3} - k_2 \frac{\partial^3 u}{\partial\alpha\partial\beta^2} \right) \right] + \left[ k_1 \frac{\partial v}{\partial\beta} - c^2 \left( \mu_1 + 3k_2 \right) \frac{\partial^3 v}{\partial\alpha^2\partial\beta} \right]$$
$$+ \left[ \left( 1 + \frac{1}{c^2} \right) k_1 + \frac{\partial^4 w}{\partial\alpha^4} + k_1 \frac{\partial^4 w}{\partial\beta^4} + 2k_1 \frac{\partial^2 w}{\partial\beta^2} + \left( 2\mu_1 + 4k_2 \right) \frac{\partial^4 w}{\partial\alpha^2\partial\beta^2} \right]$$
$$= \frac{\rho h R^2}{K} \frac{\partial^2 w}{\partial t^2}$$

(3.30)

### 3.2.4  Modal deformation displacements

One of the most important matters to apply the wave propagation approach in an appropriate way is to describe the components of displacement and rotation as suitable analytical functions which have the capability to predict the behavior of the structure with better convergence and higher accuracy. In the literature, many types of displacement functions have been implemented with various degrees of success. In the present work, to approximate the vibrational mode shapes of the shell corresponding to the various boundary conditions. This technique separates the functions of independent variables. The modal displacement fields for cylindrical shell structures can be written in the following forms:

$$u(\alpha, \beta, t) = A_m U(\alpha) \cos n\beta e^{i\omega t}$$
$$v(\alpha, \beta, t) = B_m V(\alpha) \sin n\beta e^{i\omega t} \qquad (3.31)$$
$$w(\alpha, \beta, t) = C_m W(\alpha) \cos n\beta e^{i\omega t}$$

where $U(\alpha), V(\alpha)$ and $W(\alpha)$ stand for the axial modal dependence in the longitudinal, tangential and radial directions respectively. Here $A_m, B_m$ and $C_m$ are three vibration amplitude coefficients in the three directions. The circumferential wave number is denoted by $n$ and $\omega$ indicates natural angular frequency. This frequency is associated with fundamental frequency $f$ through the following formula: $f = \omega/2\pi$.

$$A_m \frac{\partial^2 u}{\partial \alpha^2} - k_2 n^2 \left(1 + c^2\right) A_m U(\alpha) + n \left(\mu_1 + k_2\right) B_m \frac{\partial v}{\partial \alpha} + \mu_1 C_m \frac{\partial w}{\partial \alpha}$$
$$- n^2 C_m c^2 k_2 \frac{\partial w}{\partial \alpha} - C_m \frac{\partial^3 w}{\partial \alpha^3} = -\frac{\omega^2 \rho h R^2}{K} A_m U(\alpha)$$

$$(3.32)$$

$$- n A_m \left(\mu_1 + k_2\right) \frac{\partial u}{\partial \alpha} + k_2 \left(1 + 3c^2\right) B_m \frac{\partial^2 v}{\partial \alpha^2} - n^2 k_1 B_m V(\alpha)$$
$$- n k_1 C_m W(\alpha) + n c^2 \left(\mu_1 + 3k_2\right) C_m \frac{\partial^2 w}{\partial \alpha^2} = -\frac{\omega^2 \rho h R^2}{K} B_m V(\alpha)$$

$$(3.33)$$

$$\mu_1 A_m \frac{\partial u}{\partial \alpha} - c^2 \left( A_m \frac{\partial^3 u}{\partial \alpha^3} + n^2 k_2 A_m \frac{\partial u}{\partial \alpha} \right) + n k_1 B_m V(\alpha) - n c^2 \left(\mu_1 + 3k_2\right)$$
$$B_m \frac{\partial^2 v}{\partial \alpha^2} + \left(1 + \frac{1}{c^2}\right) k_1 C_m W(\alpha) + C_m \frac{\partial^4 w}{\partial \alpha^4} + n^4 k_1 C_m W(\alpha)$$
$$- 2n^2 k_1 C_m W(\alpha) - n^2 \left(2\mu_1 + 4k_2\right) C_m \frac{\partial^2 w}{\partial \alpha^2} = -\frac{\omega^2 \rho h R^2}{K} C_m W(\alpha)$$

$$(3.34)$$

## 3.3 Waves Propagation

The subscript *m* denotes the axial wave number. The functions *U*, *V*, and *W* are specified by complex exponential functions, and their exponents are related to the axial wavenumber $k_m$ for various boundary conditions

$$\text{i.e., } U(\alpha) = V(\alpha) = W(\alpha) = e^{-ik_m \alpha} \tag{3.35}$$

$k = \pi m R/L$ is the dimensionless wave-vector along the longitudinal direction, and $v$ is the wave velocity of phase. $m$ is the number of half wave in the longitudinal direction.
Using eqn (3.35). Eqn (3.32)–(3.34) takes the form

$$\left[-k_m^2 - n^2 k_2 \left(1 + c^2\right)\right] A_m + \left[n \left(\mu_1 + k_2\right) - ik_m\right] B_m$$
$$+ \left[-\mu_1 ik_m + n^2 c^2 k_2 ik_m - ik_m^3\right] C_m = -\frac{\omega^2 \rho h R^2}{K} A_m$$

$$(3.36)$$

$$[n \left( \mu_1 + k_2 \right) k_m] A_m - \left[ k_2 \left( 1 + 3c^2 \right) k_m^2 + n^2 k_1 \right] B_m$$

$$+ \left[ -nk_1 - nc^2 \left( \mu_1 + 3k_2 \right) k_m^2 \right] C_m = -\frac{\omega^2 \rho h R^2}{K} B_m \qquad (3.37)$$

$$\left[ -\mu_1 i k_m - c^2 \left( i k_m^3 - n^2 k_2 i k_m \right) \right] A_m + \left[ nk_1 + nc^2 \mu_1 k_2 k_m^2 \right] B_m$$

$$+ \left[ \left( 1 + \frac{1}{c^2} \right) k_1 + k_m^3 + n^4 k_1 - 2n^2 k_1 + n^2 \left( 2\mu_1 + 4k_2 \right) k_m^2 \right] C_m$$

$$= -\frac{\omega^2 \rho h R^2}{K} C_m$$

$$(3.38)$$

After the arrangement of terms, the above system of the equation is transformed into matrix notation, and an eigenvalue problem is produced to designate the vibration frequency equation for SWCNTs:

$$\begin{bmatrix} A_{11} & A_{12} & A_{13} \\ A_{21} & A_{22} & A_{23} \\ A_{31} & A_{32} & A_{33} \end{bmatrix} \begin{pmatrix} A_m \\ B_m \\ C_m \end{pmatrix} = -\frac{\omega^2 \rho h R^2}{K} \begin{bmatrix} 1 & 0 & 0 \\ 0 & 1 & 0 \\ 0 & 0 & 1 \end{bmatrix} \begin{pmatrix} A_m \\ B_m \\ C_m \end{pmatrix}$$

$$(3.39)$$

where the entries $A_{ij}, (i, j = 1, 2, 3)$ are labeled in the Appendix-3.1. The eigenvalues are associated with vibration frequencies of SWCNTs and the vector $\left( A_m, B_m, C_m \right)^T$ represents their mode shapes.

## 3.4 Results and Discussion

In this chapter, the orthotropic shell model based on the wave propagation approach is developed to demonstrate the natural frequencies of armchair, zigzag, and chiral SWCNTs against stiffness ($E$). The numerical values of viscoelastic material parameters such as elastic coefficient $K$ (=111.59 ± 49.9 kg/ms$^2$) and viscous coefficient $\eta$ (=55.96 ± 38.02 kg/ms$^2$). The material constants as Poisson's ratio $\mu_2 = 0.33$, mass density per unit volume $\rho = 1.47$ g/cm$^3$, Young's modulus = 0.5–2.0 Gpa, elastic Moduli are $E_1 = 1 \times 10^9$ Pa, $E_2 = 1 \times 10^6$ Pa, the equivalent thickness $h = 2.7$ nm, shear Modulus $G = 1$ MPa, effective thickness $h_0 = 1.6$ nm [30]. The numerical model based on orthotropic shell model is validated with another model. Based on the present theory, the frequencies of single-walled carbon nanotubes are compared with literature data. The present frequencies are tabulated in Table 3.1 and linked

**Table 3.1** Lists of the frequency values for armchair (5, 5), CC-CC, C-F, and aspect ratio *L/d* of CNT with Duan et al. [19].

| L/d | Frequency (THz) | | | |
|---|---|---|---|---|
| | Clamped-free | | Clamped-clamped | |
| | Present | Duan et al. [19] | Present | Duan et al. [19] |
| 5.26 | 0.199 | 0.212 | 0.951 | 0.975 |
| 5.62 | 0.173 | 0.188 | 0.835 | 0.887 |
| 5.99 | 0.159 | 0.167 | 0.786 | 0.809 |
| 6.35 | 0.139 | 0.150 | 0.711 | 0.741 |
| 6.71 | 0.129 | 0.136 | 0.639 | 0.681 |
| 7.07 | 0.1127 | 0.123 | 0.597 | 0.628 |
| 7.44 | 0.108 | 0.112 | 0.585 | 0.580 |
| 7.8 | 0.009 | 0.102 | 0.519 | 0.538 |
| 8.16 | 0.088 | 0.094 | 0.487 | 0.500 |
| 8.52 | 0.081 | 0.086 | 0.428 | 0.465 |
| 8.89 | 0.078 | 0.08 | 0.421 | 0.434 |
| 9.25 | 0.071 | 0.074 | 0.397 | 0.406 |
| 9.61 | 0.068 | 0.069 | 0.371 | 0.380 |
| 9.98 | 0.636 | 0.640 | 0.349 | 0.357 |
| 10.34 | 0.059 | 0.060 | 0.323 | 0.336 |
| 10.7 | 0.054 | 0.056 | 0.308 | 0.316 |
| 11.06 | 0.051 | 0.053 | 0.289 | 0.299 |
| 11.43 | 0.047 | 0.049 | 0.272 | 0.282 |
| 11.79 | 0.045 | 0.046 | 0.260 | 0.267 |
| 12.15 | 0.043 | 0.044 | 0.248 | 0.253 |
| 12.52 | 0.040 | 0.041 | 0.232 | 0.240 |
| 12.88 | 0.038 | 0.039 | 0.219 | 0.228 |
| 13.24 | 0.036 | 0.037 | 0.213 | 0.217 |
| 13.6 | 0.034 | 0.035 | 0.199 | 0.206 |
| 13.97 | 0.032 | 0.033 | 0.192 | 0.197 |
| 14.33 | 0.031 | 0.032 | 0.185 | 0.188 |
| 14.69 | 0.029 | 0.030 | 0.175 | 0.179 |
| 15.05 | 0.028 | 0.029 | 0.179 | 0.172 |

with Duan et al. [19]. The molecular dynamics experiment is used to find the frequencies of carbon nanotubes [19] with two boundary conditions, namely, clamped-clamped and clamped-free. The frequencies are calculated with an aspect ratio ($L/d$ = 5.26–15.05). Also, the clamped-free frequencies are lower than clamped-clamped. Through all aspect ratios, the results are well-matched with the frequencies of Duan et al. [19]. These comparisons show that present approach can give tremendous results of armchair, zigzag, and chiral SWCNTs. Also, this comparison verifies the selection of chosen parameters.

### 3.4.1 Effect of stiffness on the vibration of armchair SWCNTs

The graphs in Figures 3.3–3.5 computed the natural frequencies (THz) with respect to stiffness $E$. The optimum number of elements for armchair $(m, n)$ is considered as (5, 5), (7, 7), and (12, 12) with simply supported-simply supported and clamped-free boundary conditions. There is no restriction on choosing the number of elements. One can choose according to shorter or longer tubes [41]. The frequencies (THz) with stiffness $E$ = 1.0, 1.5 for (5, 5) are [SS-SS $\sim$ 448.85, 645.2, C–F $\sim$ 400.85, 602.43]. The frequencies increase on increasing the stiffness $E$ of the carbon nanotubes. The frequency jumps vertically upward. These sudden jumps show that the material of the tube is very stiff during vibration. The frequency gap is small between the boundary conditions of SS-SS and C–F. The frequency behavior changes for increasing the stiffness $E$ = 1.5–3.0, and bend has occurred in the frequencies because the material is less stiff. The frequency gap increases for stiffness $E$ = 1.5–2.5, and this gap decreases and becomes narrow at $E$ = 3.0. Also, another bend is occurring from $E$ = 3.0–4.0, and the gap between the curves is much

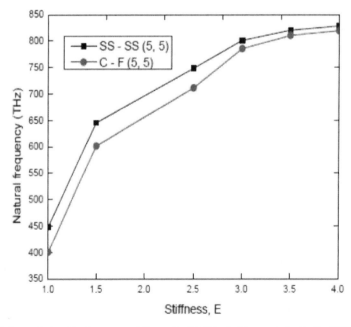

**Figure 3.3**  Relationship between stiffness $E$ (SS-SS, C–F) armchair (5, 5) SWCNTs and natural frequency.

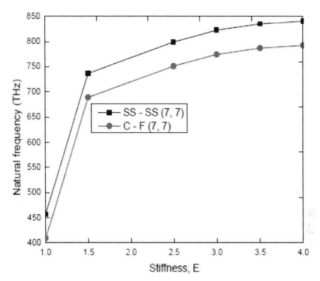

**Figure 3.4** Relationship between stiffness $E$ (SS-SS, C–F) armchair (7, 7) SWCNTs and natural frequency.

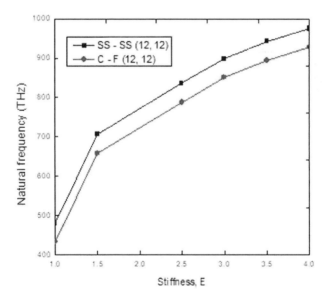

**Figure 3.5** Relationship between stiffness $E$, (SS-SS, C-F) armchair (12, 12) SWCNTs and natural frequency.

closer. The frequencies with stiffness $E = 1.0$, 1.5 for (7, 7) are [SS-SS $\sim$ 456.93, 736.79, C–F $\sim$ 408.93, 688.79]. Here, the frequency jumps suddenly more than armchair (5, 5) due to increased indices (7, 7). The frequency gap is small between the two boundary conditions for $E = 1.5$–4.0, and this gap increases two times more than $E = 1.0$, 1.5. In the case of the armchair (7, 7), there exists more bend in frequencies $E = 1.5$–4.0 than armchair (5, 5). The frequencies (THz) with stiffness $E = 1.0$, 1.5 for (12, 12) are [SS-SS $\sim$ 480.74, 706.92, C–F $\sim$ 432.74, 658.92]. The frequencies decrease on decreasing the stiffness $E$ of the carbon nanotubes. The tube is so longer due to increasing the armchair indices (12, 12) and frequencies also increases on increasing the indices. It is noted that the bend in the frequencies is seen after $E = 1.5$ for two boundary conditions. The frequency gap is unique, increasing the stiffness from $E = 1.5$–4.0. In armchair (5, 5), (7, 7), and (12, 12), frequencies increase with increasing the indices and also increases with increasing the stiffness. But the increasing behavior is different for these three different indices of armchair. The frequency pattern is the same as first increasing and then bending occur. The frequency curve is same for increasing stiffness ($E = 1.0$–1.5), and the disturbance in frequencies of SS-SS and C–F is observed after bend ($E = 1.5$). A clear pattern is observed in armchair (12, 12). It can be seen that the frequencies of clamped-free boundary condition are lower than simply supported condition. The distortion is found in the clamped-free boundary condition, and consequently, this distortion is observed for short armchair tube (5, 5). In longer armchair tubes (7, 7) and (12, 12), there is no effect of this distortion.

### 3.4.2 Effect of stiffness on the vibration of zigzag SWCNTs

From Figures 3.6–3.8, the frequency of zigzag single-walled carbon nanotubes (SWCNTs) is observed against stiffness $E$ with two prescribed boundary conditions. The index of zigzag is varied as (6, 0), (9, 0), (14, 0). The complete mechanism of frequencies in all respect is seen when vibration is produced in zigzag tube with said indices. Three different graphs are shown in Figures 3.6–3.8 to understand how zigzag frequencies affected the structure of SWCNTs. The calculated [SS-SS, C–F] frequencies at $E = 1.0$ for zigzag indices (6, 0), (9, 0), (14, 0) are [411.14, 363.14], [414.74, 366.74], [419.23, 374.23], at $E = 1.5$ are [450.69, 402.69], [453.82, 405.82], [469.97, 424.97], at $E = 2.5$ are [485.89, 437.89], [493.98, 445.98], [515.11, 470.11]. As the stiffness increases, the frequency becomes surge $E = 1.0$, 1.5 for both boundary conditions. The material is stiff, and the frequency increases.

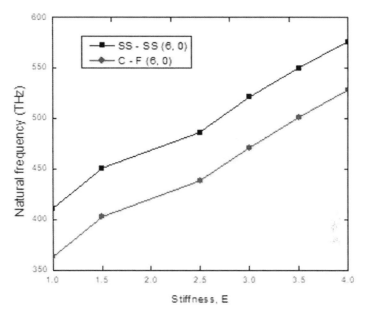

**Figure 3.6** Relationship between stiffness *E*, (SS-SS, C–F) zigzag (6, 0) SWCNTs and natural frequency

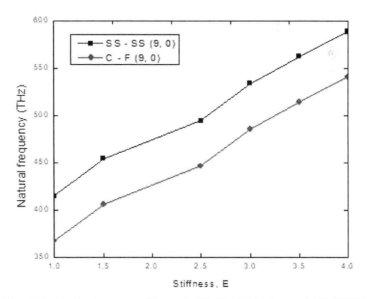

**Figure 3.7** Relationship between stiffness *E*, (SS-SS, C–F) zigzag (9, 0) SWCNTs and natural frequency.

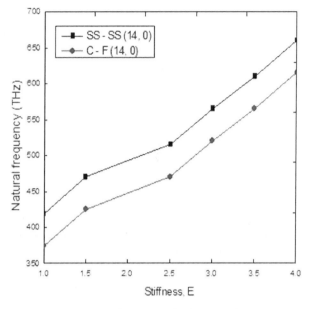

**Figure 3.8**   Relationship between stiffness *E*, (SS-SS, C–F) zigzag (14, 0) SWCNTs and natural frequency.

However, the thickness of the structure remains the same. It is noted that the frequency gap is short, and a large distance is observed for $E = 1.5$–$2.5$; after $E = 2.5$, this distance is uniform for ongoing stiffness. The calculated [SS-SS, C–F] frequencies at $E = 4.0$, for zigzag indices (6, 0), (9, 0), (14, 0) [575.52, 527.52], [588.42, 540.42], [660.38, 615.38]. It can be noted from the plot of zigzag (6, 0) the frequency turns right side and bend a bit downward $E = 1.5$–$2.5$. The material is less stiff, and the frequency bends downward. Now again, the frequency changes path after $E = 2.5$, the frequencies are growing, and the material becomes more stiff during vibration. Now in the case of zigzag (9, 0), the frequencies increase, bend and increase similar to zigzag (6, 0) reported for simply supported boundary conditions. But there is fluctuation seen in the frequency of clamped-free boundary at $E = 3.0$, and space is reduced between two boundary conditions. It is reported that overall frequencies increase on increasing the index from (6, 0) to (9, 0). Once again, the frequency marvels can be seen for zigzag (14, 0). The frequency pullover and significantly large as compared to zigzag index (6, 0)–(9, 0). A small area gap-closing is observed. After interchanging the zigzag index, various industry device components can be constructed after measurement results The structure of zigzag is beautiful and its assembly is easy to use in brick technology, welding works, and heat exchanger service life but its transformation is not easy.

Figures 3.6–3.8, the boundary of the single-walled carbon nanotubes is changed from simply supported to clamped-free condition. Therefore the stiffness of the tube is increased. Also, the deformation bending of the tube is reduced due to less stiffness of the material.

### 3.4.3 Effect of stiffness on the vibration of chiral SWCNTs

Figures 3.9–3.11 show the frequency response of chiral (6, 2), (9, 4), (12, 7) against stiffness $E$. The observed frequencies of the single-walled carbon nanotubes for different values of stiffness $E$ = 1.0, 1.5, 2.5, 3.0, 3.5, 4.0. In the past, theoretical and experimental according to stiffness is presented [42]. The single-walled carbon nanotubes can be bent easily due to their high stiffness. Also, the molecular structure of carbon nanotubes is so stiff and strongest [42]. For numerical calculation, we assume the thickness of the wall is 0.34 nm, and the cross-sectional area is $1.43 \times 10^{-18}$ m$^2$. Here, the frequencies (THz) behavior of chiral SWCNTs is studied by varying stiffness. The investigated frequencies of chiral SS-SS (6, 2) for stiffness $E$ = 1.0, 1.5, 2.5, 3.0, 3.5, 4.0 are 411.35, 674.64, 782.12, 824.23, 850.47, 876.74 and chiral SS-SS (9, 4) are 674.64, 782.12, 824.23, 850.47, 876.74, 900.1 and chiral SS-SS (12, 7) are 782.12, 820.23, 850.47, 876.74, 900.1, 925.42. The curve pattern of frequency increases with an increase of stiffness $E$ = 1.0–4.0 for chiral SS-SS (6, 2). The frequency curve of simply supported has the highest curves. The increment in frequencies is seen for stiffness $E$ = 1.0–1.5, and a bend exists upward from $E$ = 1.5–2.5. This bend shows the material is stiffer and continues till $E$ = 4.0. Also, the curve frequency decreases with stiffness decrement for chiral SS-SS (9, 4). The frequencies increase for stiffness $E$ = 1.0–1.5 for chiral SS-SS (6, 2). But there is a minute difference found for $E$ = 1.5–2.5 in comparison to chiral SS-SS (6, 2), and the pattern of curves bends a little bit. This means that material is less stiff. The frequencies proceed uniformly from $E$ = 2.5–4.0.

Moreover, frequencies for chiral SS-SS (12, 7) increase as usual for chiral SS-SS (6, 2) and (12, 7) $E$ = 1.0–1.5. Here, the phenomenon is opposite to SS-SS (6, 2) for $E$ = 1.5–2.5. The bend downward shows that the material has low stiffness. The frequencies change the path after $E$ = 2.5–4.0. It looks like a ladder and has a unique pattern, indicating that during vibration, sometimes the material is stiff as frequencies increase, and when decreasing,

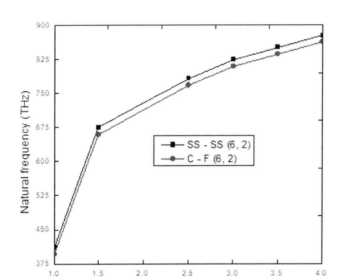

**Figure 3.9** Relationship between stiffness $E$ (SS-SS, C–F) chiral (6, 2) SWCNTs and natural frequency.

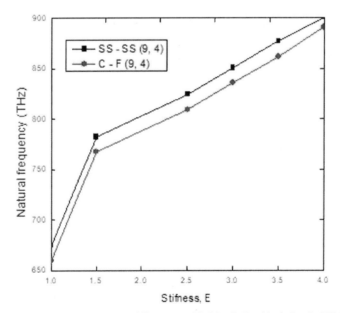

**Figure 3.10** Relationship between stiffness $E$, (SS-SS, C–F) chiral (9, 4) SWCNTs and natural frequency.

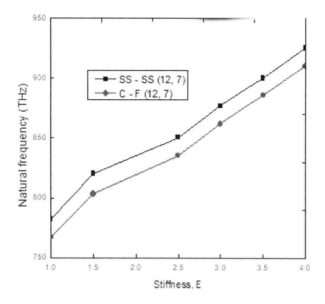

**Figure 3.11** Relationship between stiffness *E*, (SS-SS, C–F) chiral (12, 7) SWCNTs and natural frequency.

the material is less stiff. In addition, the frequencies of chiral C–F (6, 2) are 396.35, 659.64, 767.12, 809.23, 835.47, 861.74, C–F (9, 4) are 659.64, 767.12, 809.23, 835.47, 835.47, 861.74, 890.65, and C–F (12, 7) 67.12, 803.23, 835.47, 861.74, 885.65, 910.21.

With the decrease of *E* value, $f$(THz) decreases. It can be noted that similar behavior is observed for stiffness *E* = 1.0–4.0 throughout the computation for C–F (6, 2), (9, 4), and (12, 7). The simply supported frequencies are greater than that of clamped-free. In the case of chiral (6, 2) for *E* = 1.0–1.5, the gap of frequencies curves is as (12, 7) > (9, 4) > (6, 2), and gap is constant for (12, 7) and (6, 2) for *E* = 1.5–2.5. But this frequency gap for SS-SS and C–F is decreasing for *E* = 3.5–4.0.

## 3.5 Conclusions

The vibration investigation of armchair, zigzag, and chiral single-walled carbon nanotubes with the effect of stiffness has significant importance in designing and performing the structures. The validation of the present numerical approach with a large amount of exact and experimental calculations depicts the great assurance for practical engineering. Here the governing equations of the orthotropic shell model are solved by a novel wave prop-agation approach to writing the system equations in eigenform. In the case of stiffness of the material, as the stiffness increases, the frequency becomes a surge for both boundary conditions. The material is stiff than the frequency increases. However, the thickness of the structure remains the same. The pattern of curves bends a bit then, which means the material is less stiff. The simply supported frequencies are greater than that of clamped-free. This phenomina of vibration is seen for armchair, zigzag, and chairal SWCNTs in case of stiffness separately. It is necessary to see the behavior of armchair, zigzag, and chiral tubes because, in vibration, the deformations account for. So this type of deformation is a key type for the strengths of the material, and it can escape the material from a great loss. Hence, this study is a very powerful tool for organizing material in tiny instruments, sensors, and actuators.

## References

[1] Iijima, S. (1991). Helical microtubules of graphitic carbon. Nature, 354 (7) 56–58.
[2] Iijima, S., Ajayan, P. M., & Ichihashi, T. (1992). Growth model for carbon nanotubes. *Physical review letters*, *69*(21), 3100.
[3] Bethune, D. S., Kiang, C. H., De Vries, M. S., Gorman, G., Savoy, R., Vazquez, J., & Beyers, R. (1993). Cobalt-catalysed growth of carbon nanotubes with single-atomic-layer walls. Nature, 363(6430), 605-607.
[4] Blase, X., Benedict, L. X., Shirley, E. L., & Louie, S. G. (1994). Hybridization effects and metallicity in small radius carbon nan-otubes. *Physical review letters*, *72*(12), 1878.
[5] Eklund, P. C., Holden, J. M., & Jishi, R. A. (1995). Vibrational modes of carbon nanotubes; spectroscopy and theory. *Carbon*, *33*(7), 959-972.
[6] Iijima, S., Brabec, Maiti, C., A., & Bernholc. J. (1996). Structural flexibility of carbon nanotubes. Journal of Chemical Physics, 104, 2089.
[7] Kane, C. L., & Mele, E. J. (1997). Size, shape, and low energy electronic structure of carbon nanotubes. Physical Review Letters, 78(10), 1932.
[8] Krishnan, A., Dujardin, E., Ebbesen, T. W., Yianilos, P. N., & Treacy. M. M. J. (1998). Young's modulus of single-walled nanotubes. Physical Review B (Condensed Matter and Materials Physics), 58(20), 14013–14019.
[9] Nardelli, M.B., Yakobson, B.I., & Bernholc, J., (1998). Mechanism of strain release on carbon nanotubes. Phys Rev B, 57, 4277–4280.

[10] Mickelson, E. T., Huffman, C. B., Rinzler, A. G., Smalley, R. E., Hauge, R. H., & Margrave, J. L. (1998). Fluorination of single-wall carbon nanotubes. Chem Phys Lett, 296, 188–194.

[11] Miyamoto, Y., Saito, S., & Tománek, D. (2001). Electronic interwall interactions and charge redistribution in multiwall nanotubes. Physical Review B, 65(4), 041402.

[12] Harik, V. M. (2002). Mechanics of carbon nanotubes: applicability of the continuum-beam models. Comp. Mater. Sci., 24, 328–342.

[13] Flahaut, E., Bacsa, R., Peigney, A., & Laurent, C. (2003). Gram-scale CCVD synthesis of double-walled carbon nanotubes. Chemical Communications, (12), 1442-1443.

[14] Li, C. Y., & Chou, T. W. (2004). Strain and pressure sensing using single-walled carbon nanotubes. Nanotechnology, 15(11), 1493–1496.

[15] Yoon, J., Ru, C. Q., & Mioduchowski, A. (2005). Terahertz vibration of short carbon nanotubes modeled as Timoshenko beams. Journal of applied mechanics, 72(1), 10–17.

[16] Wang, Q., & Varadan, V. K. (2006). Vibration of carbon nanotubes studied using nonlocal continuum mechanics. Smart Materials and Structures, 15(2), 659.

[17] He, X. Q., Eisenberger, M., & Liew, K. M. (2006). The effect of van der Waals interaction modeling on the vibration characteristics of multi-walled carbon nanotubes. *Journal of Applied Physics*, 100(12), 124317.

[18] Wang, C. M., Tan, V. B. C., & Zhang, Y. Y. (2006). Timoshenko beam model for vibration analysis of multi-walled carbon nanotubes. *Journal of Sound and Vibration*, 294(4), 1060–1072.

[19] Duan, W. H., Wang, C. M., & Zhang, Y. Y. (2007). Calibration of nonlocal scaling effect parameter for free vibration of carbon nanotubes by molecular dynamics. Journal of applied physics, 101(2), 024305.

[20] Natsuki, T., Ni, Q. Q., & Endo, M. (2007). Wave propagation in single-and double-walled carbon nanotubes filled with fluids. Journal of Applied Physics, 101(3), 034319.

[21] Wang, L., Guo, W., & Hu, H. (2008). Group velocity of wave propagation in carbon nanotubes. In Proceedings of the Royal Society of London a: Mathematical, Physical and Engineering Sciences. 464, 2094, 1423–1438.

[22] Zhang, Y. Y., Wang, C. M., & Tan, V. B. C. (2009). Assessment of Timoshenko beam models for vibrational behavior of single-walled carbon nanotubes using molecular dynamics. *Advances in Applied Mathematics and Mechanics*, 1(1), 89-106.

[23] Elishakoff, I., & Pentaras. D. (2009). Fundamental natural frequencies of double-walled carbon nanotubes. Journal of Sound and Vibration, 322, 652–664.

[24] Khademolhosseini, F., Rajapakse, R. K. N. D., & Nojeh, A. (2010). Torsional buckling of carbon nanotubes based on nonlocal elasticity shell models. Computational materials science, 48(4), 736-742.

[25] Madihav, M. H., Jiang, L. Y., & Sun, X. (2011). Nonlinear vibration of a double-walled carbon nanotube embedded in a polymer matrix. Physica E: Low-dimensional Systems and Nanostructures, 43(10), 1813–1819.

[26] Khademolhosseini, F., Phani, A. S., Nojeh, A., & Rajapakse, N. (2012 Basirjafari, S., Esmaeilzadeh Khadem, S., & Malekfar, R. (2013). Validation of shell theory for modeling the radial breathing mode of a single-walled carbon nanotube. Int. J. Eng. Trans. A, 26(4), 447-454.

[27] Basirjafari, S., Esmaeilzadeh Khadem, S., & Malekfar, R. (2013). Validation of shell theory for modeling the radial breathing mode of a single-walled carbon nanotube. Int. J. Eng. Trans. A, 26(4), 447-454.

[28] Mikhasev, G. (2014). On localized modes of free vibrations of single-walled carbon nanotubes embedded in nonhomogeneous elastic medium. ZAMM-Journal of Applied Mathematics and Mechanics/Zeitschrift für Angewandte Mathematik und Mechanik, 94(1-2), 130-141.

[29] Chemi, A., Heireche, H., Zidour, M., Rakrak, K., & Bousahla, A. A. (2015). Critical buckling load of chiral double-walled carbon nanotube using non-local theory elasticity. Advances in Nano Research, 3(4), 193.

[30] Wang, J., and Gao, Y. (2016), "Nonlocal orthotropic shell model applied on wave propagation in microtubules", *Applied Mathematical Modelling*, **40**(11-12), 5731-5744.

[31] Soltani, P., Saberian, J., & Bahramian, R. (2016). Nonlinear vibration analysis of single-walled carbon nanotube with shell model based on the nonlocal elasticity theory. Journal of Computational and Nonlinear Dynamics, 11(1), 011002.

[32] Bouadi, A., Bousahla, A.A., Houari, M.S.A., Heireche, H., and Tounsi, A., (2018), "A new nonlocal HSDT for analysis of stability of single layer graphene sheet", Advances in Nano Research, 6(2), 147-162.

[33] Fatahi-Vajari, A., Azimzadeh, Z., & Hussain, M. (2019). Nonlinear coupled axial–torsional vibration of single-walled carbon nanotubes using homotopy perturbation method. Micro & Nano Letters, 14(14), 1366-1371.

[34] Malikan, M., & Eremeyev, V. A. (2020). Post-critical buckling of truncated conical carbon nanotubes considering surface effects embedding in a nonlinear Winkler substrate using the Rayleigh-Ritz method. Materials Research Express, 7(2), 025005.

[35] Pourasghar, A., Yang, W., Brigham, J., & Chen, Z. (2021). Nonlocal thermoelasticity: Transient heat conduction effects on the linear and nonlinear vibration of single-walled carbon nanotubes. Mechanics Based Design of Structures and Machines, 1-17.

[36] Ghasemi, S. E., & Gouran, S. (2022). Nonlinear Analysis on Flow-Induced Vibration of Single-Walled Carbon Nanotubes Employing Analytical Methods. International Journal of Structural Stability and Dynamics, 2250115.

[37] Zhang, X. M., Liu, G. R., & Lam, K. Y. (2001). Coupled vibration analysis of fluid-filled cylindrical shells using the wave propagation approach, *Applied Acoustics*, 62, 229–243.

[38] S. Fluegge , Stresses in Shells, 1, 2nd ed., Springer, Berlin, 1973 (1973).

[39] R. Zou , C. Foster , Simple solution for buckling of orthotropic circular cylindrical shells, Thin-walled Struct. 22 (1995) 143–158.

[40] Y. Gao , L. An , A nonlocal elastic anisotropic shell model for microtubule buckling behaviors in cytoplasm, Phys. E: Low-dimens. Syst. Nanostruct. 42 (2010) 2406–2415.

[41] Thostenson, E. T., Ren, Z., & Chou, T. W. (2001). Advances in the science and technology of carbon nanotubes and their composites: a review. *Composites science and technology*, *61*(13), 1899-1912.

[42] Strozzi, M., Manevitch, L. I., Pellicano, F., Smirnov, V. V., & Shepelev, D. S. (2014). Low-frequency linear vibrations of single-walled carbon nanotubes: Analytical and numerical models. *Journal of Sound and Vibration*, *333*(13), 2936-2957.

# 4

# Donnell Shell Theory Formulation – Single-walled Carbon Nanotubes: Frequency Assessment via Height-to-Diameter Ratios

## Abstract

Theoretical formulations based on Donnell's shell theory are developed for vibration characteristics of armchair, zigzag, and chiral single-walled carbon nanotubes. The governing equation of motion and boundary conditions using the wave propagation approach is written in the form of eigenvalue to extract the frequencies of CNTs. This modified model has less complication and has been compared with the earlier methods. The effect of height-to-diameter ratios, boundary conditions for armchair (4, 4), (11, 11), zigzag (6, 0), (13, 0), and chiral (7, 4), (12, 6) on the frequencies are calculated. Based on Donnell's shell theory, it is observed that the increase in the height-to-diameter ratios, the frequencies increase for all tubes (armchair, zigzag, and chiral). These variations of frequencies are seen with clamped-clamped and clamped-free end conditions. The frequency pattern with two prescribed boundary conditions seems parallel for overall height-to-diameter ratio values. It is also seen that the frequency curves with changing values of height-to-diameter ratio of C–F boundary condition are the lowest outcomes. The presence of chirality of armchair, zigzag, and chiral have no effect on the vibration frequencies because the frequencies increase on increasing the height-to-diameter ratios. But it has momentus effect on the deformation of the tubes. The present frequencies are compared with beam element, Timoshenko beam model, and molecular dynamic (MD) simulations with properly chosen parameters. The comparison shows that the present model can lead to appropriate results for the prediction of frequencies.

**Keywords:** Donnell's shell theory, armchair, zigzag, and chiral SWCNTs, clamped-clamped and clamped-free, frequency pattern.

## 4.1 Introduction

The animation of carbon nanotubes with small diameters is described by Iijima [1]. Nowadays, carbon nanotube application is widely used in many nanomaterials [2, 3]. It was in 1993 that Iijima and Ichihashi [4] and Bethune et al. [5] found an effective method to syntheses single-walled carbon nanotubes, independently. This decade saw a large amount of growth in the field of nanostructure, such as micro-electronics, energy, biological applications, and mechanical [6, 7]. To examine the geometrical feathers of SWCNTs, the vibration behavior is needed [8, 9]. Nardelli et al. [10] constructed the ideal CNTs with hexagonal ring and sp2 hybridized carbon atoms. An example of hexagone-heptagone is Stone-Wales defects comprising five membered rings. WenXing et al. [11] studied the interaction of short and long-range CNTs based on molecular dynamics (MD) simulation. The second-generation reactive empirical bond order (REBO) potential is used for the simulation of armchair, zigzag, and chiral carbon nanotubes. Zhang et al. [12] studied the transverse vibrations of the carbon nanotubes using a nonlocal Euler-Bernoulli beam model for the first time. Natsuki et al. [13] demonstrated the vibration of SWCNTs using Flügge shell model, and the embedded CNTs are described in an elastic medium with the Winkler model. The frequencies are analyzed and investigated based on FSM to find the effect of vibrational modes and nanotube parameter. Zhang et al. [14] analyzed the elastic buckling of carbon nanotubes with the energy method. The effect of van der Waals forces on hydrostatic pressure is studied. The DWCNTs are supposed to be thin in the surrounding medium. The natural frequencies of single- and double-walled CNTs obtained by nonlocal elastic beam model [15] considered various estimation of dimensions and length of both tubes, and analytical simulations were performed to quantify small-scale effects. The determined results based on nonlocal continuum model were found in good accordance with published experimental ones. Vibration characteristics of SWCNTs and DWCNTs were conducted using a flexible shell model [16]. Benzair et al. [17] used TBM for the vibration of SWCNTs. The transverse shear effect is investigated with length-to-diameter ratios. Murmu and Pradhan [18] investigated the vibrational frequencies with different modes along temperature change using nonlocal small-scale effects. Conversely, the nonlocal frequencies are comparatively lower for length scale coefficient and soft elastic medium with embedded carbon nanotube. It is also found that the frequencies of the nonlocal model at different stages of temperature are higher than the nonlocal with the same temperature. Zhange et al. [19] investigated the vibration characteristics of SWCNTs based on Timoshenko beam models. The effect of aspect ratio versus natural frequency is seen.

Pradhan [20] investigated the vibrational frequencies with different modes along temperature change using nonlocal small-scale effects. On the other side, for the length scale coefficient and soft elastic medium with embedded carbon nanotube, the nonlocal frequencies are comparatively lower. Kulathunga et al. [21] described the influence of axial buckling of SWCNTs for various configurations of vacancy defects of atomic structures in different thermal environments using MD simulations. They found that there is a significant reduction in buckling strain of SWCNTs by missing a single atom. Arghavan and Singh [22] presented the free and forced vibration of SWCNTs based on numerical technique. The armchair and zigzag structure is observed throughout the study. The clamped and free boundary condition is applied on the structure. The natural frequencies and modes shapes of such structures are observed. Mahdavi et al. [23] suggested the vibration of DWCNTs with van der Waals forces for both adjacent and medium tubes. They derived a relation between TBM and EBM through the harmonic balance method. They observed for the first time the bi-directional vibration of SWCNTs with the Timoshenko beam model. Rayleigh and TBM were considered, and the effects of electromagnetic fields were studied on them [24]. Ansari and Hemmatnezhad [25] studied the vibrations of double-walled carbon nanotubes using nonlinear finite element analysis, where Von-Kamran type nonlinear strain-displacement relationships were employed to constraint the ends of carbon nanotubes to move axially. Das et al. [26] studied the nonlocal theories for the in-extensional vibration of SWCNTs. The in-extensional mode frequency is treated by the positive strain gradient theory (SGT) with circumferential wave number. Kiani [27] presented a comprehensive discussion on the vibration and unstableness of single-walled CNTs in a 3-D area of magnetic influence. Bending frequency, related velocity, and buckling of CNTs also included with a diverse magnetic field. The implementation of wave propagation is substantial for the investigation of nano-sized structures to formulate with distinguishing theories. In the application of the method, the eigenvalue system is established aided by the axial mode function in the form of matrix. Computer software MATLAB has been operated to extract the frequency vibration of double-walled CNTs. Besseghier et al. [28] presented the nonlinear vibration of zigzag SWCNTs based on the Winkler-type model (WTM). The energy-equivalent model was used for the derivation of the general equation. Ehyaei and Daman [29] practiced nonlocal Timoshenko beam model to examine embedded double-walled

CNTs in the elastic medium at the outset imperfection. Free natural frequencies of double-walled CNTs were assisted by naiver exact solutions. Natural frequency showed an increasing trend due to curvature amplitude. Ebrahimi and Mahmoodi [30] presented the static analysis of SWCNTs and the vibration of CNTs using Eringen's beam theory. The bending moment and function of strain were performed with different boundary conditions. Recently some material researchers explained the axial vibration analysis, buckling response, and effect of nanofluid with cracked nanorod using nonlocal theory with coupled nonlinear equation [31].

Bensattalah et al. [32] implemented the TBM to study the vibration response of SWCNTs. The coupled solution is obtained to see the frequency influence of chirality and wave mode of SWCNTs. Also, the effect of the small-scale coefficient with aspect ratio is discussed. Jena et al. [33] carried the shear deformation beam theory to observe the vibration of SWCNTs. This novel theory has fewer variables as compared to the deformation theory. These tubes are placed in the magnetic field. Miyashiro et al. [34] studied the behavior of DNA-wrapped SWCNTs. This study is helpful for bio-sensing in lipid membranes and for many applications such as drug delivery and bio-imaging. Ebrahimi [35] investigated the complexities in the vibration of CNTs based on nonlocal theory using simply supported conditions. The governing equations of motions are solved by Rung–Kutta method (RKM).

The present study was performed on the vibration of armchair, zigzag, and chiral SWCNTs to provide the natural frequencies against height-to-diameter ratios under clamped-clamped (C–C) and clamped-free boundary conditions. The results presented by this model are in good agreement with the results of the beam element, Timoshenko beam model, MD simulations, proving that the parameters are correctly chosen. Furthermore, eigensolutions of the frequency equation have been determined by writing in MATLAB coding. The present comprehensive results may also be helpful in designing the fundamental frequencies of nanostructures.

## 4.2 Mathematical Formulation

### 4.2.1 Equation of motion using shell model

Carbon nanotubes have two configurations, single-walled carbon nanotubes (SWCNTs) and multi-walled nanotubes (MWCNTs), and due to radii, the structure differs. When the graphene sheet is rolled once, then it becomes a cylinder with an end cape and it is treated as SWCNTs. The schematic configuration of graphene sheets with SWCNTs is presented in Figure 4.1.

Figure 4.1 represents a side lateral view armchair, zigzag, and chiral tubes. The shape of the SWCNT is correlated to the hollow cylinder due to its geometrical structure, as shown in Figure 4.2. To study the vibration of SWCNTs, the governing equation of He et al. [36] for the cylindrical shell is engaged.

A SWCNT with geometrical structure has been shown in Figure 4.1, where $Eh, R, v, \rho h, \varepsilon, \kappa, \zeta$ and $t$, denoted as in-plane rigidity, the radius, Poisson ratio, mass density per unit lateral area of the tube, longitudinal,

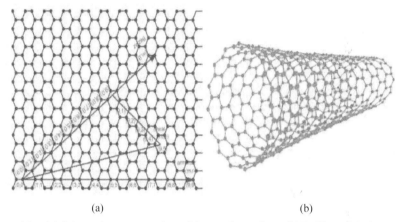

(a)                                             (b)

**Figure 4.1** (**a**) Schematic representation of the graphene sheet. (**b**) Rolling of the hexagonal sheet.

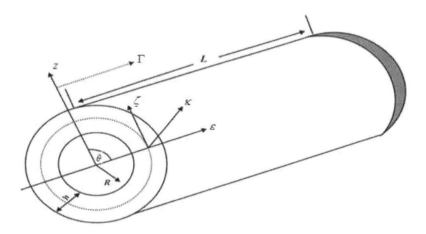

**Figure 4.2** Geometry of SWCNTs.

circumferential, and radial displacements and time, respectively and the effective bending stiffness is denoted by $D = Eh^3/12\left(1 - v^2\right)$. The vibration equations [36] of SWCNTs considering geometric parameter in the form of PDE are articulated as:

$$\frac{\partial^2 \varepsilon}{\partial \Gamma^2} + \frac{1 - v}{2R^2}\frac{\partial^2 \varepsilon}{\partial Y^2} + \frac{1 + V}{2R}\frac{\partial^2 \kappa}{\partial \Gamma \partial Y} - \frac{V}{R}\frac{\partial \zeta}{\partial \Gamma} = \frac{\left(1 - v^2\right)\rho h}{Eh}\frac{\partial^2 \varepsilon}{\partial t^2} \quad (4.1)$$

$$\frac{1 + v}{2R}\frac{\partial^2 \varepsilon}{\partial \Gamma \partial Y} + \frac{1 - v}{2}\frac{\partial^2 x}{\partial \Gamma^2} + \frac{1}{R^2}\frac{\partial^2 x}{\partial Y^2} - \frac{1}{R^2}\frac{\partial \zeta}{\partial Y} = \frac{\left(1 - v^2\right)\rho h}{Eh}\frac{\partial^2 \kappa}{\partial t^2}$$
$$(4.2)$$

$$\frac{v}{R}\frac{\partial \varepsilon}{\partial \Gamma} + \frac{1}{R^2}\frac{\partial \kappa}{\partial \Gamma} - \left(\frac{1}{R^2} + \frac{\left(1 - v^2\right)}{Eh}D\left(\frac{\partial^4 \zeta}{\partial \Gamma^4} + 2\cdot\frac{1}{R^2}\frac{\partial^4 \zeta}{\partial \Gamma^2 \partial Y^2}\right.\right.$$
$$\left.\left. + \frac{1}{R^4}\frac{\partial^4 \zeta}{\partial Y^4}\right)\right) = \frac{\left(1 - v^2\right)\rho h}{Eh}\frac{\partial^2 \zeta}{\partial t^2} \quad (4.3)$$

## 4.3 Numerical Approach

### 4.3.1 Modal displacement form

In the present problem, displacement functions are adopted to eliminate the independent spatial and temporal variables. For this, the following modal form is assumed as

$$\varepsilon(\Gamma, \Upsilon, t) = j_\tau \sin(n\Upsilon)\exp\left[-i\left(l_c\Gamma - \omega t\right)\right] \quad (4.4)$$
$$\kappa(\Gamma, \Upsilon, t) = k_\tau \mathrm{Cos}(n\Upsilon)\exp\left[-i\left(l_c\Gamma - \omega t\right)\right] \quad (4.5)$$
$$\zeta(\Gamma, \Upsilon, t) = r_\tau \sin(n\Upsilon)\exp\left[-i\left(l_c\Gamma - \omega t\right)\right] \quad (4.6)$$

The coefficients $(j_\tau, k_\tau, r_\tau)$ represents the amplitudes of vibration, respectively, in the $s$, $\kappa$ and $z$ directions. Here $\omega, m, n$, are denoted by angular frequency, axial half and circumferential wave numbers, respectively and $l_c$ is designated for the axial wave number specified for boundary conditions. The formula for fundamental natural frequency is $f = \omega/2\pi$. Here the deformation $\varepsilon(\Gamma, \Upsilon, t)$, $\zeta(\Gamma, \Upsilon, t)$ are, respectively, in the direction of axial, tangential, and radial the tube and apply the product method for splitting the space and temporal variable. The simplified SWCNT equation can be written together with their partial derivatives after substituting eqn (4.4)–(4.6) into eqn (4.1)–(4.3)

$$\left(l^2{}_c + \frac{1-v}{2R^2}n^2\right)j_\tau + \left(il_c\frac{1+v}{2R}n\right)k_\tau + (il_c\frac{v}{R})r_\tau = \frac{(1-v^2)\rho h}{Eh}\omega^2 j_\tau$$

(4.7)

$$\left(-n\frac{1+v}{2R}il_c\right)j_\tau + \left(\frac{1-v^2}{2}l_c + \frac{n^2}{R^2}\right)k_\tau + \left(\frac{n}{R^2}\right)r_\tau = \frac{(1-v^2)\rho h}{Eh}\omega^2 k_\tau$$

(4.8)

$$\left(-\frac{v}{R}il_c\right)j_\tau + \left(\frac{n}{R^2}\right)k_\tau + \left(\frac{1}{R^2} + \frac{(1-v^2}{Eh}D\left(l^4{}_c + 2\frac{1}{R^2}l^2{}_cn^2 + \frac{n^4}{R^4}\right)\right)$$

$$r_\tau = \frac{(1-v^2)\rho h}{Eh}\omega^2 j_\tau$$

(4.9)

The matrix equation to represent the frequencies, after the arrangement of terms can be written as.

$$\begin{bmatrix} \Delta_{11} & \Delta_{12} & \Delta_{13} \\ \Delta_{21} & \Delta_{22} & \Delta_{23} \\ \Delta_{31} & \Delta_{32} & \Delta_{33} \end{bmatrix} \begin{pmatrix} j_\tau \\ k_\tau \\ r_\tau \end{pmatrix} = \frac{(1-v^2)\rho h}{Eh}\omega^2 \begin{bmatrix} 1 & 0 & 0 \\ 0 & 1 & 0 \\ 0 & 0 & \end{bmatrix} \begin{pmatrix} j_\tau \\ k_\tau \\ r_\tau \end{pmatrix}$$

(4.10)

The expression $\Delta_{ij}$ 's has been given in Appendix 4.1. $(j_\tau, k_\tau, r_\tau)$ in the form of non-zero solution to generate SWCNT vibration frequency and correlation mode and the root of the equation provides the frequency. As evidenced in order to adjust the fundamental frequency of vibrations of the CNTs, the lowest root corresponds to minimum frequency w.r.t wave number $m, n$. The eigenvectors are related to shell mode shapes and the eigenvalues of the problem correspond to shell frequencies.

## 4.4 Numerical Results

### 4.4.1 Parametric study

To investigate the vibration characteristics of armchair, zigzag and chiral SWCNTs, the ratio of height-to-diameter is taken into account with two different boundary conditions (C–C and C–F). The parameters are selected from earlier literature as Poisson ratio $v = 0.19$ [14], the mass density $\rho = 2.3$ g/cm$^3$ [17], and the equaivalent thickness of CNTs $h = 0.34$ nm [19]. The Young's modulus of the SWCNTs is obtained by MD simulation [19] (also see Chapter 2).

## 4.4.2 Validation

In order to know the vibration characteristics of armchair (4, 4), (11, 11), zigzag (6, 0), (13, 0), and chiral (7, 4), (12, 6) SWCNTs using DST based on WPA, in the initial step, the accuracy is performed by validating the results. Tables 4.1 and 4.2 show the comparison of present results with previously published data (MD simulation and TBM) [19] with clamped-free and clamped-clamped boundary conditions. The natural frequencies versus aspect ratio are seen to be well matched. Another comparison is made for frequencies of armchair (6, 6) and zigzag (8, 0) with the results

**Table 4.1**   Natural frequencies of C-F SWCNTs: Comparison of present model with TBM and MDs [19].

| L/d | Frequencies (THz) | | |
| --- | --- | --- | --- |
| | Present | MD simulation [19] | TBM [19] |
| 4.67 | 0.22290 | 0.23193 | 0.26357 |
| 6.47 | 0.11376 | 0.12872 | 0.14134 |
| 7.55 | 0.08735 | 0.09766 | 0.10491 |
| 8.28 | 0.06987 | 0.07935 | 0.0876 |
| 10.07 | 0.04389 | 0.05493 | 0.0596 |
| 13.69 | 0.02853 | 0.03052 | 0.03248 |
| 17.3 | 0.01711 | 0.01831 | 0.02041 |
| 20.89 | 0.01217 | 0.01381 | 0.01401 |
| 24.5 | 0.00897 | 0.00916 | 0.0102 |
| 28.07 | 0.0061 | 0.0069 | 0.00777 |
| 31.64 | 0.0059 | 0.0061 | 0.00612 |
| 35.34 | 0.0043 | 0.0045 | 0.00491 |

**Table 4.2**   Natural frequencies of C-C SWCNTs: Comparison of present model with TBM and MDs

| L/d | Frequencies (THz) | | |
| --- | --- | --- | --- |
| | Present | MD simulation [19] | TBM [19] |
| 4.86 | 1.05732 | 1.06812 | 1.14229 |
| 6.67 | 0.58871 | 0.64697 | 0.69764 |
| 8.47 | 0.38973 | 0.43335 | 0.46628 |
| 10.26 | 0.27996 | 0.30518 | 0.33214 |
| 13.89 | 0.17329 | 0.18311 | 0.18999 |
| 17.49 | 0.10735 | 0.11597 | 0.12255 |
| 21.06 | 0.06435 | 0.07629 | 0.08555 |
| 24.66 | 0.05128 | 0.05798 | 0.06285 |
| 28.85 | 0.04382 | 0.04578 | 0.04796 |
| 31.85 | 0.03432 | 0.03662 | 0.03800 |
| 35.53 | 0.02991 | 0.03052 | 0.03062 |

of Arghavan and Singh [22], as shown in Tables 4.3 and 4.4. The frequencies for different mode numbers with two boundary conditions having fixed diameter and length. It is seen that the frequencies are in closed agreement. After validation, it appears an excellent accuracy. Results based on the present model provide a powerful technique for investigating the behavior of SWCNTs.

The schema of the applied boundary condition is shown in Figure 4.3.

**Table 4.3** Natural frequencies of armchair (6, 6) SWCNTs with diameter = 0.814 nm and length = 5.66 nm: comparison of the present model with Arghavan and Singh [22].

| Model no. | Frequencies (THz) | | | |
| | Clamped-clamped | | Clamped-free | |
| | Present | Arghavan and Singh [22] (2011) | Present | Arghavan and Singh [22] |
|---|---|---|---|---|
| 1 | 0.5216 | 0.5374 | 0.1027 | 0.1039 |
| 2 | 0.5312 | 0.5401 | 0.1039 | 0.1042 |
| 3 | 1.2098 | 1.2155 | 0.5512 | 0.5565 |
| 4 | 1.2327 | 1.2486 | 0.5539 | 0.5592 |
| 5 | 1.2498 | 1.2671 | 0.6224 | 0.6255 |
| 6 | 1.7786 | 1.7866 | 0.9360 | 0.9371 |
| 7 | 1.7756 | 1.7872 | 1.3090 | 1.3180 |
| 8 | 1.8498 | 1.8526 | 1.3348 | 1.3358 |
| 9 | 1.9299 | 1.9367 | 1.7517 | 1.7522 |
| 10 | 1.933 | 1.936 | 1.7539 | 1.7542 |

**Table 4.4** Natural frequencies of zigzag (8, 0) SWCNTs with diameter = 0.626 nm and length = 5.826 nm: comparison of the present model with Arghavan and Singh [22].

| Model no. | Frequencies (THz) | | | |
| | Clamped-clamped | | Clamped-free | |
| | Present | Arghavan and Singh [22] | Present | Arghavan and Singh [22] |
|---|---|---|---|---|
| 1 | 0.3786 | 0.4323 | 0.0723 | 0.0746 |
| 2 | 0.4197 | 0.4323 | 0.0739 | 0.0746 |
| 3 | 1.0487 | 1.0582 | 0.4287 | 0.4316 |
| 4 | 1.0519 | 1.0582 | 0.4299 | 0.4316 |
| 5 | 1.2579 | 1.2661 | 0.6183 | 0.6296 |
| 6 | 1.8071 | 1.8194 | 0.9021 | 0.9049 |
| 7 | 1.8281 | 1.8391 | 1.0888 | 1.0917 |
| 8 | 1.8270 | 1.8391 | 1.0901 | 1.0917 |
| 9 | 2.5198 | 2.5284 | 1.8784 | 1.8868 |
| 10 | 2.6823 | 2.6993 | 1.9022 | 1.9031 |

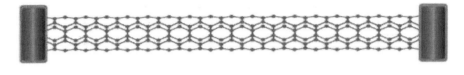

(a)

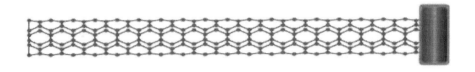

(b)

**Figure 4.3**   Schema of boundary conditions (**a**) C–C. (**b**) C–F.

### 4.4.3 Effect of height–diameter ratio on the vibration of armchair single-walled carbon nanotubes

For better understanding, the variation of natural frequencies is carried out keeping the same referenced parameters with BCs (C–C & C–F) for aspect ratio on different modes of vibration. Figures 4.4 and 4.5 show the frequency response of armchair SWCNTs versus the ratio of height-to-diameter $h/d$ with two different boundary conditions. The frequencies at a ratio of height-to-diameter for C–C [(4, 4), (11, 11)] $f$(THz) $\sim$ (42.389, 42.526, 42.751, 43.062, 43.457, 43.934, 44.491, 45.123, 45.829, 46.605), (22.389, 22.526, 22.751, 23.062, 23.457, 23.934, 24.491, 25.123, 25.829, 26.605). It can take into account that the armchair SWCNTs (4, 4) and (11, 11), with C–C condition, have the prominent and highest frequencies. As $h/d$ is enhanced, the frequency goes up. The frequencies increase slowly for the short armchair tube (4, 4). The fundamental natural frequencies increase robustly for all calculated values of ratio ($h/d$ = 0.001–0.010) in the case of armchair (11, 11), respectively. Once again, the frequencies outcomes with clamped-free boundary conditions as C–F [(4, 4), (11, 11)] $f$(THz) $\sim$ (42.304, 65.44, 92.451, 120.86, 149.91, 179.29, 207.86, 237.56, 267.34, 298.18), (47.304, 70.44, 97.451, 125.86, 154.91, 184.29, 213.86, 243.56, 273.34, 303.18), respectively. The frequency increases as the $h/d$ grows up. The sketched frequency curves at ($h/d$ = 0.001–0.010) for armchair (11, 11) are higher than

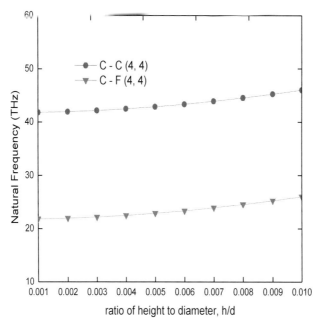

**Figure 4.4** Response of frequency response of armchair SWCNTs (4, 4), versus *h/d* for C–C, C–F BC's.

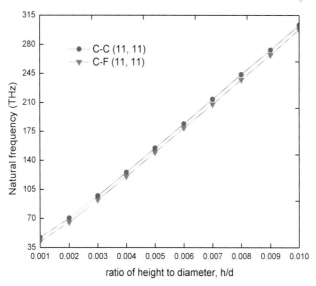

**Figure 4.5** Response of frequency response of armchair SWCNTs (11, 11) versus *h/d* for C–C, C–F BC's.

armchair (4, 4) frequency curves. The frequency pattern with two boundary conditions seems to be parallel for armchair (11, 11) and insignificant. The symmetric frequency for BCs is C–C > C–F, and the frequency continuation can be elaborated mathematically as armchair (4, 4) < armchair (11, 11). It is observed that the frequencies C–C and C–F are highly visible and have large distances between the curves for armchair (4, 4). It is also concluded that the frequency curves with changing the values of $h/d$ of the C–F boundary condition are the lowest outcomes.

### 4.4.4 Effect of height–diameter ratio on the vibration of zigzag single-walled carbon nanotubes

Figures 4.6 and 4.7 show the frequency analysis of zigzag (6, 0) and (13, 0) against the ratio of height-to-diameter with a length of 2.4 nm. These variations of frequencies are drawn with clamped-clamped and clamped-free end conditions. The investigated values of frequencies with different boundary conditions are at $h/d$ = 0.001 as (6, 0) (C–C) $f \sim$ 6.2815; (6, 0) (C–F) $f \sim$ 3.2815, at $h/d$ = 0.002 as (6, 0) (C–C) $f \sim$ 11.723; (6, 0) (C–F) $f \sim$ 8.7230, at $h/d$ = 0.003 as (6, 0) (C–C) $f \sim$ 17.305; (6, 0) (C–F) $f \sim$ 14.305, at $h/d$ = 0.004 as (6, 0) (C–C), = $f \sim$ 22.923; (6, 0) (C–F) $f \sim$ 19.923, at $h/d$ = 0.005 as (6, 0) (C–C) $f \sim$ 28.557; (6, 0) (C–F) $f \sim$ 25.557, at $h/d$ =

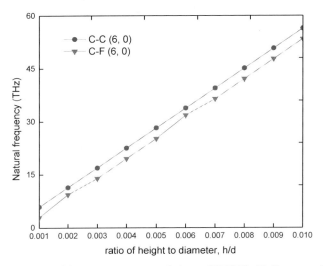

**Figure 4.6**   Response of frequency response of zigzag SWCNTs (6, 0), versus $h/d$ for C–C, C–F BC's.

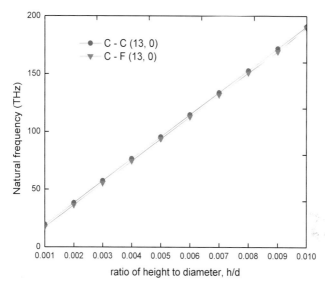

**Figure 4.7**  Response of frequency response of zigzag SWCNTs (13, 0), versus *h/d* for C–C, C–F BC's.

0.006 as (6, 0) (C–C) $f \sim$ 34.198; (6, 0) (C–F) $f \sim$ 31.198, at *h/d* = 0.007 as (6, 0) (C–C) $f \sim$ 39.843; (6, 0) (C–F) $f \sim$ 36.843, at *h/d* = 0.008 as (6, 0) (C–C) $f \sim$ 45.491; (6, 0) (C–F) $f \sim$ 42.491, at *h/d* = 0.009 as (6, 0) (C–C) $f \sim$ 51.141; (6, 0) (C–F) $f \sim$ 48.141, at *h/d* = 0.010 as (6, 0) (C–C) $f \sim$ 56.791; (6, 0) (C–F) $f \sim$ 53.791. Both C–C and C–F frequencies increases as the ratio of height to diameter is increased. It is noted that frequencies of zigzag (6, 0) increases symmetrically except for two points (*h/d* = 0.002, 0.005). At these two points (*h/d* = 0.002, 0.005), the C–F frequencies rust up to the C–C frequencies, and the gap between the frequency of these two curves is insignificant. As shown by this figure, the boundary conditions C–C have the highest frequency curves. The investigated values of frequencies with different boundary conditions are at *h/d* = 0.001 as (13, 0) (C–C) $f \sim$ 19.67; (13, 0) (C–F) $f \sim$ 17.67, at *h/d* = 0.002 as (13, 0) (C–C) $f \sim$ 38.546; (13, 0) (C–F) $f \sim$ 36.546, at *h/d* = 0.0013 as (13, 0) (C–C) $f \sim$ 57.573; (13, 0) (C–F) $f \sim$ 55.573, at *h/d* = 0.004 as (13, 0) (C–C) $f \sim$ 76.639; (13, 0) (C–F) $f \sim$ 74.639, at *h/d* = 0.005 as (13, 0) (C–C) $f \sim$ 95.72; (13, 0) (C–F) $f \sim$ 93.72, at *h/d* = 0.006 as (13, 0) (C–C) $f \sim$ 114.8; (13, 0) (C–F) $f \sim$ 112.8, at *h/d* = 0.007 as (13, 0) (C–C) $f \sim$ 133.9; (13, 0) (C–F) $f \sim$ 131.9, at *h/d* = 0.008 as (13, 0) (C–C) $f \sim$ 0. 152.99; (13, 0) (C–F) $f \sim$ 150.99, at *h/d* = 0.009 as

(13, 0) (C–C)$f \sim$ 172.09; (13, 0) (C–F)$f \sim$169.69, at $h/d$ = 0.010 as (13, 0) (C–C)$f \sim$ 191.19; (13, 0) (C–F)$f \sim$ 189.19. The frequencies are pronounced on increasing ratio of height-to-diameter. The gap between frequency curves is little significant as compared to zigzag (6, 0) for all values of ratio $h/d$. The frequency pattern with two prescribed boundary conditions seems to be parallel for overall values of $h/d$ (= 0.001–0.010) except for one point $h/d$ = 0.010. The C–C and C–F frequencies are overlapped for $h/d$ = 0.010. It is also concluded that the frequency curves with changing the values of $h/d$ of C–F boundary condition are the lowest outcomes. It is noted that increasing the indices of zigzag (6, 0) to (13, 0), the frequencies increases. With higher index, the frequencies will be higher.

### 4.4.5 Effect of height–diameter ratio on the vibration of chiral single-walled carbon nanotubes

Figures 4.8–4.9 shows the chiral $(7, 4), (12, 6)$ single walled carbon nanotubes as obtained from is shown DST. Here, a nice piece of work on vibration of chiral is investigated tubes. Now the fully analysis of frequencies of said chiral indices. The investigated values of frequencies with different boundary conditions are C $-$ C$[(7, 4) \sim (12, 6)]$ as $[(79.589) \sim (58.589)]$ at

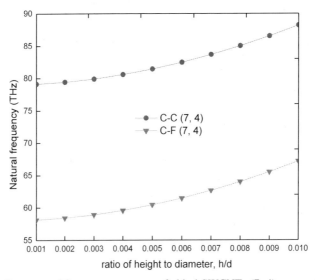

**Figure 4.8**   Response of frequency response of chiral SWCNTs (7, 4), versus $h/d$ for C–C, C–F BC's.

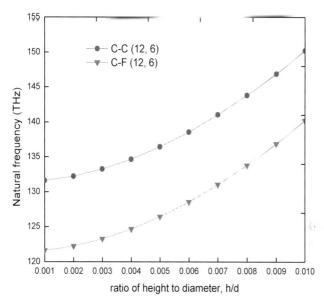

**Figure 4.9**   Response of frequency response of chiral SWCNTs (12, 6), versus *h/d* for C–C, C–F BC's.

$h/d = 0.001, \mathrm{C} - \mathrm{C}[(7,4) \sim (12,6)]$ as $[(79.885) \sim (58.885)]$ at $h/d = 0.002, \mathrm{C} - \mathrm{C}[(7,4) \sim (12,6)]$ as $[(80.371) \sim (59.371)]$ at $h/d = 0.003,$ $\mathrm{C-C}[(7,4) \sim (12,6)]$ as $[(81.042) \sim (60.042)]$ at $h/d = 0.004, \mathrm{C-C}[(7,4) \sim (12,6)]$ as $[(81.895) \sim (60.895)]$ at $h/d = 0.005, \mathrm{C} - \mathrm{C}[(7,4) \sim (12,6)]$ as $[(82.923) \sim (61.923)]$ at $h/d = 0.006, \mathrm{C-C}[(7,4) \sim (12,6)]$ as $[(84.12) \sim (63.12)]$ at $h/d = 0.007, \mathrm{C} - \mathrm{C}[(7,4) \sim (12,6)]$ as $[(85.48) \sim (64.48)]$ at $h/d = 0.008, \mathrm{C} \text{-} \mathrm{C} [(7,4) \sim (12,6)]$ as $[(86.993) \sim (65.993)]$ at $h/d = 0.009, \mathrm{C} - \mathrm{C}[(7,4) \sim (12,6)]$ as $[(88.653) \sim (67.65)]$ at $h/d = 0.010.$

As $h/d$ is increased, the backward and forward frequencies (Hz) get upward for chiral $(7,4)$ and $(12,6)$. The investigated values of frequencies with different boundary conditions are $C - F[(7,4) \sim (12,6)]$ as $[(131.99) \sim (121.99)]$ at $h/d = 0.001, \mathrm{C} - \mathrm{F}[(7,4) \sim (12,6)]$ as $[(132.61) \sim (122.61)]$ at $h/d = 0.002, \mathrm{C} - \mathrm{F}[(7,4) \sim (12,6)]$ as $[(133.63) \sim (123.63)]$ at $h/d = 0.003, \mathrm{C} - \mathrm{F}[(7,4) \sim (12,6)]$ as $[(135.04) \sim (125.04)]$ at $h/d = 0.004, \ C - F[(7,4) \sim (12,6)]$ as $[(136.81) \sim (126.81)]$ at $h/d = 0.005,$ $\mathrm{C-F}[(7,4) \sim (12,6)]$ as $[(138.95) \sim (128.95)]$ at $h/d = 0.006, \mathrm{C-F}[(7,4) \sim (12,6)]$ as $[(141.41) \sim (131.41)]$ at $h/d = 0.007, \ C - F[(7,4) \sim (12,6)]$ as $[(144.19) \sim (134.19)]$ at $h/d = 0.008, \ \mathrm{C} - \mathrm{F}[(7,4) \sim (12,6)]$ as

$[(147.26) \sim (137.26)]$ at $h/d = 0.009$, $\mathrm{C-F}[(7,4) \sim (12,6)]$ as $[(150.59)$ $\sim (140.59)]$ at $h/d = 0.010$. Interestingly, as $h/d$ is enhanced, the frequency (THz) grows for C-F condition. The C-C frequencies are higher than those corresponding C-F condition. For shorter tube and shorter chiral index, the frequency displacement between the curves of $\mathrm{C-C}$ and $\mathrm{C-F}$ boundary condition is large. As the index is increased from $(7,4)$ to $(12,6)$, the displacement of frequency curves is smaller. As frequency proceed upward on increasing $h/d$ for $(12,6)$, there is decrement seen in the displacement of frequency curve.

## 4.5 Conclusions

This chapter investigated the vibration of armchair, zigzag, and chiral single-walled carbon nanotubes based on Donnell shell theory. This model contains both the effect of boundary conditions and height-to-diameter ratios, and the wave propagation approach is engaged to decrease the governing equations in eigenform. This eigenform is solved through MATLAB software to obtain the fundamental frequencies of SWCNTs. The effect of boundary conditions and height-to-diameter ratios for frequency behavior is discussed. The frequency pattern with two boundary conditions seems to be parallel for armchair, zigzag, and chiral. The frequencies are pronounced on increasing ratio of height-to-diameter. With a higher index, the frequencies will be higher. The C–C frequencies are higher than those corresponding C–F conditions. For shorter tube and chiral index, the frequency displacement between the curves of C–C and C–F boundary conditions is large. The author expects this frequency analysis for high frequencies in fascinating electromagnetic devices.

## References

[1] Iijima, S. (1991). Helical microtubules of graphitic carbon. Nature, 354(7), 56–58.

[2] Sawada, S. I., & Hamada, N. (1992). Energetics of carbon nanotubes. *Solid State Communications*, 83(11), 917-919.

[3] Ajayan, P. M. (1993). Capillarity-induced filling of carbon nanotubes. *Nature*, 361(6410), 333-334.

[4] S. Iijima and T. Ichihashi, "Single-shell carbon nanotubes of 1-nm diameter," 1993.

[5] D. Bethune, C. Klang, M. De Vries, G. Gorman, R. Savoy, J. Vazquez, and R. Beyers, "Cobalt-catalysed growth of carbon nanotubes with single-atomic-layer walls," 1993.

[6] Planeix, J. M., Coustel, N., Coq, B., Brotons, V., Kumbhar, P. S., Dutartre, R., ... & Ajayan, P. M. (1994). Application of carbon nanotubes as supports in heterogeneous catalysis. *Journal of the American Chemical Society*, *116*(17), 7935-7936.

[7] Dresselhaus, M. S., Dresselhaus, G., & Saito, R. (1995). Physics of carbon nanotubes. *Carbon*, *33*(7), 883-891.

[8] Treacy, M. J., Ebbesen, T. W., & Gibson, J. M. (1996). Exceptionally high Young's modulus observed for individual carbon nanotubes. *nature*, *381*(6584), 678-680.

[9] Rao, A. M., Richter, E., Bandow, S., Chase, B., Eklund, P. C., Williams, K. A., ... & Dresselhaus, M. S. (1997). Diameter-selective Raman scattering from vibrational modes in carbon nanotubes. *Science*, *275*(5297), 187-191.

[10] Nardelli, M.B., Yakobson, B.I., & Bernholc, J., (1998). Mechanism of strain release on carbon nanotubes. Phys Rev B, 57, 4277–4280.

[11] WenXing, B., ChangChun, Z., & WanZhao, C. (2004). Simulation of Young's modulus of single-walled carbon nanotubes by molecular dynamics. *Physica B: Condensed Matter*, *352*(1-4), 156-163.

[12] Zhang, Y. Q., Liu, G. R., & Xie, X. Y. (2005). Free transverse vibrations of double-walled carbon nanotubes using a theory of nonlocal elasticity. Physical Review B, 71(19), 195404.

[13] Natsuki, T., Endo, M., & Tsuda, H. (2006). Vibration analysis of embedded carbon nanotubes using wave propagation approach. Journal of applied physics, 99(3), 034311

[14] Zhang, Y. Q., Liu, G. R., Qiang, H. F., & Li, G. Y. (2006). Investigation of buckling of double-walled carbon nanotubes embedded in an elastic medium using the energy method. *International journal of mechanical sciences*, *48*(1), 53-61.

[15] Wang, Q., & Varadan, V. K. (2006). Vibration of carbon nanotubes studied using nonlocal continuum mechanics. Smart Materials and Structures, 15(2), 659.

[16] Yan, Y., He, X. Q., Zhang, L. X. & Wang, Q. (2007). Flow-induced instability of double-walled carbon nanotubes based on an elastic shell model. Journal of Applied physics, 102(4), 044307.

[17] Benzair, A., Tounsi, A., Besseghier, A., Heireche, H., Moulay, N., & Boumia, L. (2008). The thermal effect on vibration of single-walled

carbon nanotubes using nonlocal Timoshenko beam theory. *Journal of Physics D: Applied Physics, 41*(22), 225404.

[18] Murmu, T. & S. C. Pradhan, (2009). Thermo-mechanical vibration of a single-walled carbon nanotube embedded in an elastic medium based on nonlocal elasticity theory. Computational Materials Science, 46(4), 854–859.

[19] Zhang, Y. Y., Wang, C. M., & Tan, V. B. C. (2009). Assessment of Timoshenko beam models for vibrational behavior of single-walled carbon nanotubes using molecular dynamics. *Advances in Applied Mathematics and Mechanics, 1*(1), 89-106.

[20] Pradhan, S. C., & Phadikar, J. K. (2009). Small scale effect on vibration of embedded multilayered graphene sheets based on nonlocal continuum models. Physics Letters A, 373(11), 1062-1069.

[21] Kulathunga, D. D. T. K., Ang, K. K., & Reddy, J. N. (2010). Molecular dynamics analysis on buckling of defective carbon nanotubes. Journal of Physics: Condensed Matter, 22(34), 34530.

[22] Arghavan, S., & Singh, A. V. (2011). On the vibrations of single-walled carbon nanotubes. *Journal of Sound and Vibration, 330*(13), 3102-3122.

[23] Mahdavi, M. H., Jiang, L. Y., & Sun, X. (2011). Nonlinear vibration of a double-walled carbon nanotube embedded in a polymer matrix. Physica E: Low-dimensional Systems and Nanostructures, 43(10), 1813–1819.

[24] Kiani, K., (2012). Transverse wave propagation in elastically confined single-walled carbon nanotubes subjected to longitudinal magnetic fields using nonlocal elasticity models. Physica E: Low-Dimensional Systems and Nanostructures, 45, 86–96.

[25] Ansari, R., & Hemmatnezhad, M. (2012). Nonlinear finite element analysis for vibrations of double-walled carbon nanotubes. Nonlinear Dynamics, 67(1), 373-383.

[26] Das, S. L., Mandal, T., & Gupta, S. S. (2013). Inextensional vibration of zig-zag single-walled carbon nanotubes using nonlocal elasticity theories. International Journal of Solids and Structures, 50(18), 2792–2797.

[27] Kiani, K. (2014). Vibration and instability of a single-walled carbon nanotube in a three dimensional magnetic field. Journal of Physics and Chemistry of Solids, 75(1), 15–22.

[28] Besseghier, A., Heireche, H., Bousahla, A.A., Tounsi, A., and Benzair, A., (2015), "Nonlinear vibration properties of a zigzag single-walled carbon nanotube embedded in a polymer matrix", Advances in nano research, 3(1), 029.

[29] Fhyaei, J., & Daman, M. (2017). Free vibration analysis of double walled carbon nanotubes embedded in an elastic medium with initial imperfection. Advances in Nano Research, 5(2), 179.

[30] Ebrahimi F., and Mahmoodi, F., (2018), "Vibration analysis of carbon nanotubes with multiple cracks in thermal environment", Adv Nano Res, 6(1), 57-80. DOI:

[31] Akbaş, Ş. D. (2019), "Axially Forced Vibration Analysis of Cracked a Nanorod", Journal of Computational Applied Mechanics, 50(1), 63-68.

[32] Bensattalah, T., Zidour, M., & Daouadji, T. H. (2019). A new nonlocal beam model for free vibration analysis of chiral single-walled carbon nanotubes. Compos. Mater. Eng, 1(1), 21-31.

[33] Jena, S. K., Chakraverty, S., Malikan, M., & Mohammad-Sedighi, H. (2020). Hygro-magnetic vibration of the single-walled carbon nanotube with nonlinear temperature distribution based on a modified beam theory and nonlocal strain gradient model. International Journal of Applied Mechanics, 12(05), 2050054.

[34] Miyashiro, D., Hamano, R., Taira, H., & Umemura, K. (2021). Analysis of vibration behavior in single strand DNA-wrapped single-walled carbon nanotubes adhered to lipid membranes. Forces in Mechanics, 2, 100008.

[35] Ebrahimi, R. (2022). Chaotic vibrations of carbon nanotubes subjected to a traversing force considering nonlocal elasticity theory. Proceedings of the Institution of Mechanical Engineers, Part N: Journal of Nanomaterials, Nanoengineering and Nanosystems, 236(1-2), 31-40.

[36] He, X. Q., Eisenberger, M., & Liew, K. M. (2006). The effect of van der Waals interaction modeling on the vibration characteristics of multi-walled carbon nanotubes. *Journal of Applied Physics*, 100(12), 124317.

# 5

# Impact of Poisson's Ratios on the Vibration of Single-walled Carbon Nanotubes: Prediction of Frequencies through Galerkin's Technique

## Abstract

An analysis of the Poisson's ratio influence of SWCNTs based on Sander's shell theory (SST) is carried out. The effect of Poisson's ratio, boundary conditions, and different armchair, chiral, and zigzag, and chiralities of carbon nanotubes SWCNTs is discussed and studied. An original method of Sander's shell theory is proposed with two different boundary conditions such as simply supported and clamped simply supported in order to investigate the natural frequency in detail of these chiralities. A change of natural frequency for different Poisson's ratios of SWCNTs is conducted. Moreover, this model predicts phenomena of frequencies of chiralities of SWCNTs. The frequency response is investigated for different indices of chiral, zigzag, and armchair SWCNTs (5, 3), (8, 2), (11, 7), and (7, 0), (11, 0), (15, 0), and (6, 6), (8, 8), (10, 10), respectively. With the decrease in ratios of Poisson, the frequency increases. Poisson's ratio directly measures the deformation in the material. A high Poisson's ratio denotes that the material exhibits large elastic deformation. Due to this deformation, frequencies of carbon nanotubes increase. The frequency value increases with the increase of indices of single-walled carbon nanotubes. The prescribed boundary conditions used are simply supported and clamped simply supported. The Timoshenko beam model (TBM) is used to compare the results. The present method should serve as benchmark results as it is in good agreement with the results of other models, with a slightly different value of the natural frequencies.

**Keywords:** Poisson's ratio, chiralities, Sander's shell theory, clamped simply supported, natural frequencies.

## 5.1 Introduction

Carbon nanotubes have been considerable importance due to a number of publications. The carbon nanotubes have grown very quickly in the last decades, since their discovery [1]. There are many applications of carbon nanotubes in energy storage, optical, composite materials, electronics, sensors, oxidation, doping, and field-emission electron [2–5]. The geometrical properties such as high aspect ratio, length-to-diameter ratio, and chirality make CNTs more effectively. In order to investigate these geometrical parameters, vibration of carbon nanotubes is remarkable interesting [6]. Vibrations of CNTs have a great influence in nanomechanical devices such as sensors, clocks, oscillators, and charge detectors and in many evaluation processes [7].

Kwon and Tomanek [8] investigated the atomic morphology of multi-walled CNTs and disclosed the fact of double-walled CNTs being significant component in the formation of multi-walled tubes. Leung and Kuang [9] conducted a numerical study of MWCNTs based on Flügge's theory. Van der Waals interactions and hydrostatic pressure of MWCNTs have no significant effect on natural frequency. He et al. [10] investigated the vibration frequencies of MWCNTs based on Donnell shell theory and the radial frequencies of CNTs are obtained by these explicit formulas. The results showed that frequencies are small and can be ignored due to van der Waals effect. Domination of rotary inertia was described by using Timoshenko beam theory through vibrational behavior of multi-walled CNTs. Rabczuk et al. [11] exhibited mesh-free approach to analyze the capacity of mesh-free fluid model with vigorous fracture of fluid-filled cylindrical shell that is jolted by penetrated projectile. The formulation of fluid shell model was based on Kirchhoff–Love theory. Sun and Liu [12] performed frequency analysis of MWCNTs with radial pressure and initial axial force. To obtain the natural frequencies of DWCNTs, the explicit expression was used with initial axial stress. They found that with increasing the internal pressure, the natural frequencies increase and the frequencies decrease with the decrease of external pressure. Moreover, the natural frequencies increase by increasing the tensile stress and decrease by increasing the compression stress. They also discussed the stress for lower tubes and insensitivity to initial stress for inter-tubes. Liew and Wang [13] discussed wave dispersion for the highest modes of single- and double-walled CNTs. At one time, thin and thick Love and Cooper Naghdi shell theories have been used to study shear and inertia significance. The feasibility and effective use of elastic models have been explained by comparison of outputs. Hu et al. [14] reported a study on the transverse and torsion waves based on nonlocal shell model for single-walled and double-walled CNTs. Prominent computational competence and accuracy make nonlocal models an attractive choice for further advancements in field. Natsuki et al. [15] determined the frequencies of DWCNTs and assessed the effects of van der Waals force using EBM and Winkler spring model. Demir et al. [16] studied the vibration analysis of CNTs using Discrete singular convolution (DSC) method. The numerical accurate results are presented. Yang et al. [17] presented the frequency spectrum of CNTs based on nonlocal theory and geometric theory. The significance of nonlocal parameter on tube radius and height is deliberated explicitly. MD simulation has been used to estimate the vibrational frequency of zigzag (5, 0), (8, 0), (9, 0), (11, 0) single-walled CNTs with certain end conditions. Ghavanloo and Fazelzadeh [18] exhibited the remarkable mechanical phenomena on vibration of chiral SWCNTs using the Flügge's shell theory. The SWCNT's torsional, radial, and longitudinal frequencies were also studied. It is concluded that the present problem can be used for the atomistic and material properties and also in major engineering and scientific interests.

Fang [19] utilized nonlocal theory to examine forced nonlinear vibrational aspects of double-walled CNTs. The governing equations have been formulated with the help of EBM and Hamilton principle. He considered nonlinearity of Von Karman geometry and nonlinear van der Waal forces. Noncoaxial vibration amplitudes were observed to be small subjected to both geometric nonlinearity and nonlinear van der Waal forces. Ansari and Arash [20] investigated vibrations of DWCNTs based on nonlocal theory using DQM. The mechanical behavior of DWCNTs with geometrical parameters layer-wise boundary conditions and small-scale factors is fully investigated. Chawis et al. [21] investigated the effects of different geometrical boundary conditions and tube chirality are considered. The numerical results with low aspect ratios are in good agreement with classical solution. Furthermore, in this study, the first order-ten modes for boundary conditions and different aspect ratios and repeated natural frequencies are also highlighted. Liang et al. [22] investigated the buckling and vibration of simply supported beam. The Euler beam model based on gradient theory is used for solving the governing equations analytically. Simply supported beam was used for the free vibration of beam. On increasing the thickness of the beam, the Poisson's effect increases.

Ansari and Rouhi [23] studied stability of SWCNTs under axial load and demonstrated the small-scale effects of lengths based on RRM. The axial buckling of armchair (8, 8) SWCNTs is found with various boundary conditions by applying the molecular dynamic simulation and by adjusting the nonlocal parameter with bending rigidity and in-plane stiffness to predict the results of MD simulations. Torkaman et al. [24] conducted the analysis on vibrations and steadiness of rotating single-walled CNTs premised on nonlocal elasticity theory and assumptions considered from Love theory. Exact and authentic results have been established through nonlocal model indicating the influence of rotation rates and role of elasticity for rotating devices. Faroughi and Shaat [25] investigated the mechanical property (Poisson's ratio) for deformation materials. The majority of the material has features of positive Poisson's ratio and some auxetic materials have negative Poisson's ratios. Eltaher et al. [26] focused on the characterization of single-walled carbon nanotubes to evaluate the Young's modulus using tensile test. The natural frequency equation is derived with FEM. The outcomes of the results gave new ideas for research in micro-structures. Su and Cho [27] used TBM for free vibration of SWCNTs. The effects of atomic structure with different boundary conditions in elastic medium are examined in detail. The slenderness ratio is affected with higher modes of SWCNTs. Selim [28] presented the vibration of SWCNTs for the influence of surface irregularity. The Donnell's model is used to see the parabolic form which is made by the surface irregularity.

The present chapter is concerned for vibration characteristics of armchair, zigzag, and chiral SWCNTs with the use of Sander's shell theory to analyze the effects of Poisson's ratios. The suggested technique to investigate the solution of fundamental eigen relations is Galerkin's method, which is an eminent and effectual method to form the fundamental frequency equations. It is carefully observed from the literature, and no information is seen regarding the present developed model where such problem has been considered, so it became an incentive to conduct the current study. The frequency value of armchair, zigzag, and chiral tubes indicates that the addition of Poisson effect increases the effective stiffness of single-walled carbon nanotubes.

## 5.2 Structural Analysis and Modeling

### 5.2.1 Theoretical formulation

Studying single-walled carbon nanotubes presented a prodigious insight on rolling the piece of graphene sheet which is in the form of honey comb and look like a cylinder having very small diameter. The schematic illustration is shown in Figures 5.1 and 5.2. In this chapter, the Sander's shell theory for

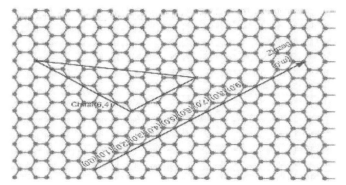

**Figure 5.1** Schematic illustration of graphene sheet.

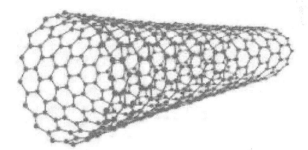

**Figure 5.2** Schematic illustration of SWCNTs.

the vibration of single-walled carbon nanotubes, using Galerkin's technique (GLT), is developed. The geometrical parameters are sketched in Figure 2.1 (see Chapter 2). The material parameters are Young's modulus ($E$), the Poisson's ratio ($v$), and mass density ($\rho$).

## 5.3 Application of Sander's Shell Theory

In the last two centuries, the intensive study is done by the researchers to find the accurate results related to material of cylindrical shape. These studies are based on the suppositions and assumptions in theoretical form before its practical use. Kirchhoff presented a very important assumption for such type of materials which is "Normal to the un-deformed middle surface of a shell remain normal to it, undergoes no change in length during deformation." Many researches used the Kirchhoff assumption with its modified form and achieved fruitful results. Sander's shell theory based on Kirchhoff assumption

[29] is applied in the present study due to its simple mathematical formulation and to explore new results of cylindrical nature especially single-walled carbon nanotubes.

The expressions for strain and curvature displacement relationship are written as

$$
\left\{ \begin{array}{c} e_1 \\ e_2 \\ \gamma \end{array} \right\} = \left\{ \begin{array}{c} \frac{\partial u}{\partial x} \\ \frac{1}{R}\left(\frac{\partial v}{\partial \psi} + w\right) \\ \left(\frac{\partial v}{\partial x} + \frac{1}{R}\frac{\partial u}{\partial \psi}\right) \end{array} \right\} + z \left\{ \begin{array}{c} -\frac{\partial^2 w}{\partial x^2} \\ -\frac{1}{R^2}\left(\frac{\partial^2 w}{\partial \theta^2} - \frac{\partial v}{\partial \theta}\right) \\ -\frac{2}{R}\left(\frac{\partial^2 w}{\partial x \partial \theta} - \frac{3}{4}\frac{\partial v}{\partial x} + \frac{1}{4R}\frac{\partial u}{\partial \theta}\right) \end{array} \right\}
$$

(5.1)

The stress and moments results can be obtained by [33]

$$
N = (N_x, N_\theta, N_{x\theta})^T \, dz = \int_{-\frac{h}{2}}^{\frac{h}{2}} (\sigma_x, \sigma_\psi, \sigma_{x\psi})^T \, dz
$$

$$
M = (M_x, M_\theta, M_{x\theta})^T \, dz = \int_{-\frac{h}{2}}^{\frac{h}{2}} (\sigma_x, \sigma_\psi, \sigma_{x\psi})^T \, z dz
$$

(5.2)

From the relations (5.2), a relationship among the forces, moments, and the surface strains and surface curvatures is written as

$$
\begin{bmatrix} N_x \\ N_\psi \\ N_{x\psi} \\ M_x \\ M_\psi \\ M_{x\psi} \end{bmatrix} = \begin{bmatrix} A_{11} & A_{12} & 0 & B_{11} & B_{12} & 0 \\ A_{12} & A_{22} & 0 & B_{12} & B_{22} & 0 \\ 0 & 0 & A_{66} & 0 & 0 & B_{66} \\ B_{11} & B_{12} & 0 & D_{11} & D_{12} & 0 \\ B_{12} & B_{22} & 0 & D_{12} & D_{22} & 0 \\ 0 & 0 & B_{66} & 0 & 0 & D_{66} \end{bmatrix} \begin{bmatrix} e_1 \\ e_2 \\ \gamma \\ \kappa_1 \\ \kappa_2 \\ 2\tau \end{bmatrix}
$$

(5.3)

where $A_{ij}$, $B_{ij}$, and $D_{ij}$ ($i, j$ = 1, 2, and 6) denote the membrane, coupling, and flexural, respectively, and are expressed as

$$
A_{ij} = \int_{-\frac{h}{2}}^{\frac{h}{2}} Q_{ij} dz
$$

(5.4a)

$$
B_{ij} = \int_{-\frac{h}{2}}^{\frac{h}{2}} Q_{ij} z dz
$$

(5.4b)

$$D_{ij} = \int_{\frac{-h}{2}}^{\frac{h}{2}} Q_j z^2 dz \qquad (5.4c)$$

$[Q]$ stands for the material reduced stiffness for a cylindrical shell. Coupling stiffness, $B_{ij}$'s, vanishes for a cylindrical shell structured from isotropic materials where they exist for heterogeneous and an-isotropic materials such as laminated, functionally graded materials. This is owing to the un-symmetric distribution of material around the shell middle surface.

The strain energy, $S$ is given by (see Chapter 2)

$$S = \frac{R}{2} \int_0^L \int_0^{2\pi} \Big[ A_{11}e_1^2 + A_{22}e_2^2 + 2A_{12}e_1e_2 + A_{66}\gamma^2 + 2B_{11}e_1k_1$$

$$+ 2B_{12}e_1k_2 + 2B_{12}e_2k_1 + 2B_{22}e_2k_2 + 4B_{66}\gamma\tau + D_{11}k_1^2 + D_{22}k_2^2$$

$$+ 2D_{12}k_1k_2 + 4D_{66}\tau^2 \Big] d\theta dx \qquad (5.5)$$

Also for a cylindrical shell, its kinetic energy expression represented by $T$, is given as

$$T = \frac{R}{2} \int_0^L \int_0^{2\pi} \rho_T \left[ \left( \frac{\partial u}{\partial t} \right)^2 + \left( \frac{\partial v}{\partial t} \right)^2 + \left( \frac{\partial w}{\partial t} \right)^2 \right] d\theta dx \qquad (5.6)$$

where $\rho_T$ is defined as mass density in a unit length and is expressed as:

$$\rho_T = \int_{\frac{-h}{2}}^{\frac{h}{2}} \rho dz \qquad (5.7)$$

and $\rho$ is the shell mass density.

Using the strain and curvature displacement relations from eqn (5.1) in the expression (5.5) for the shell strain energy, the following form of this energy is written as:

$$S = \frac{R}{2} \int_0^L \int_0^{2\pi} \left[ A_{11} \left( \frac{\partial u}{\partial x} \right)^2 + \frac{A_{22}}{R^2} \left( \frac{\partial v}{\partial \theta} + w \right)^2 + \frac{2A_{12}}{R} \left( \frac{\partial u}{\partial x} \right) \right.$$

$$
\left(\frac{\partial v}{\partial \theta} + w\right) + A_{66}\left(\frac{\partial v}{\partial x} + \frac{1}{R}\frac{\partial u}{\partial \theta}\right)^2 - 2B_{11}\left(\frac{\partial u}{\partial x}\right)\left(\frac{\partial^2 w}{\partial x^2}\right)
$$

$$
- \frac{2B_{12}}{R^2}\left(\frac{\partial u}{\partial x}\right)\left(\frac{\partial^2 w}{\partial \theta^2} - \frac{\partial v}{\partial \theta}\right) - \frac{2B_{12}}{R}\left(\frac{\partial v}{\partial \theta} + w\right)\left(\frac{\partial^2 w}{\partial x^2}\right)
$$

$$
- \frac{2B_{22}}{R^3}\left(\frac{\partial v}{\partial \theta} + w\right)\left(\frac{\partial^2 w}{\partial \theta^2} - \frac{\partial v}{\partial \theta}\right) - \frac{8B_{66}}{R}\left(\frac{\partial v}{\partial x} + \frac{1}{R}\frac{\partial u}{\partial \theta}\right)
$$

$$
\left(\frac{\partial^2 w}{\partial x \partial \theta} - \frac{3}{4}\frac{\partial v}{\partial x} + \frac{1}{4R}\frac{\partial u}{\partial \theta}\right) + D_{11}\left(\frac{\partial^2 w}{\partial x^2}\right)^2
$$

$$
+ \frac{D_{22}}{R^4}\left(\frac{\partial^2 w}{\partial \theta^2} - \frac{\partial v}{\partial \theta}\right)^2 + \frac{2D_{12}}{R^2}\left(\frac{\partial^2 w}{\partial x^2}\right)\left(\frac{\partial^2 w}{\partial \theta^2} - \frac{\partial v}{\partial \theta}\right)
$$

$$
- \frac{8D_{66}}{R^2}\left(\frac{\partial^2 w}{\partial x \partial \theta} - \frac{3}{4}\frac{\partial v}{\partial x} + \frac{1}{4R}\frac{\partial u}{\partial \theta}\right)^2 d\theta dx \tag{5.8}
$$

Now the shell problem is framed by Lagrangian energy functional which is difference between the shell kinetic and strain energies and is given as

$$
\Pi = T - S \tag{5.9}
$$

By substituting the values of $T$ and $S$ from eqn (5.6) and (5.8) in eqn (5.9), the obtained expression is given as

$$
\Pi = \frac{R}{2}\int_0^L \int_0^{2\pi}\left(\rho_T\left[\left(\frac{\partial u}{\partial t}\right)^2 + \left(\frac{\partial v}{\partial t}\right)^2 + \left(\frac{\partial w}{\partial t}\right)^2\right] - A_{11}\left(\frac{\partial u}{\partial x}\right)^2\right.
$$

$$
+ \frac{A_{22}}{R^2}\left(\frac{\partial v}{\partial \theta} + w\right)^2 + \frac{2A_{12}}{R}\left(\frac{\partial u}{\partial x}\right)\left(\frac{\partial v}{\partial \theta} + w\right)
$$

$$
+ A_{66}\left(\frac{\partial v}{\partial x} + \frac{1}{R}\frac{\partial u}{\partial \theta}\right)^2 - 2B_{11}\left(\frac{\partial u}{\partial x}\right)\left(\frac{\partial^2 w}{\partial x^2}\right)
$$

$$
- \frac{2B_{12}}{R^2}\left(\frac{\partial u}{\partial x}\right)\left(\frac{\partial^2 w}{\partial \theta^2} - \frac{\partial v}{\partial \theta}\right) - \frac{2B_{12}}{R}\left(\frac{\partial v}{\partial \theta} + w\right)\left(\frac{\partial^2 w}{\partial x^2}\right)
$$

$$
- \frac{2B_{22}}{R^3}\left(\frac{\partial v}{\partial \theta} + w\right)\left(\frac{\partial^2 w}{\partial \theta^2} - \frac{\partial v}{\partial \theta}\right) - \frac{8B_{66}}{R}\left(\frac{\partial v}{\partial x} + \frac{1}{R}\frac{\partial u}{\partial \theta}\right)
$$

$$\left( \frac{\partial^2 w}{\partial x \partial \theta} - \frac{3}{4} \frac{\partial v}{\partial x} + \frac{1}{4R} \frac{\partial u}{\partial \theta} \right) + D_{11} \left( \frac{\partial^2 w}{\partial x^2} \right)^2$$

$$+ \frac{D_{22}}{R^4} \left( \frac{\partial^2 w}{\partial \theta^2} - \frac{\partial v}{\partial \theta} \right)^2 + \frac{2D_{12}}{R^2} \left( \frac{\partial^2 w}{\partial x^2} \right) \left( \frac{\partial^2 w}{\partial \theta^2} - \frac{\partial v}{\partial \theta} \right)$$

$$- \frac{8D_{66}}{R^2} \left( \frac{\partial^2 w}{\partial x \partial \theta} - \frac{3}{4} \frac{\partial v}{\partial x} + \frac{1}{4R} \frac{\partial u}{\partial \theta} \right)^2 \Bigg) d\theta dx \qquad (5.10)$$

where $\Pi$ denotes the Lagrangian functional (LGF).

Hamilton's variational principle is applied to the energy functional (5.9), i.e.,

$$\int_{t_1}^{t_2} \delta(\Pi) dt = \int_{t_1}^{t_2} \delta(S - J) dt \qquad (5.11)$$

where $t_2 - t_1$ describes a very smalltime interval.

## 5.4 Derivation of Shell Governing Equations

Calculus of variations process is applied to the integral terms to derive the Euler–Lagrange equations. Hamilton's variational principle is a process in which the variations in the variables are assumed to be zero. This principle is implemented to the Lagrangian functional which is an integral expression to derive the shell governing equations in the compact operator form as

$$L_{11}u + L_{12}v + L_{13}w = \rho_T \frac{\partial^2 u}{\partial t^2} \qquad (5.12a)$$

$$L_{21}u + L_{22}v + L_{23}w = \rho_T \frac{\partial^2 v}{\partial t^2} \qquad (5.12b)$$

$$L_{31}u + L_{32}v + L_{33}w = \rho_T \frac{\partial^2 w}{\partial t^2} \qquad (5.12c)$$

where the operators $L_{ij}$, $s(i, j = 1, 2, 3)$ are the differential operators involving the variable $x$ and $\psi$. the expression for the differential $L_{ij}$, $s(i, j = 1, 2, 3)$ in the shell motion equations, full form of these equations is given as:

$$A_{11} \frac{\partial^2 u}{\partial x^2} + \frac{A_{66}}{R^2} \frac{\partial^2 u}{\partial \theta^2} + \left( \frac{A_{12} + A_{66}}{R} + \frac{B_{12} + 2B_{66}}{R^2} \right) \frac{\partial^2 v}{\partial x \partial \theta} + \frac{A_{12}}{R} \frac{\partial w}{\partial x}$$

$$- B_{11} \frac{\partial^3 w}{\partial x^3} - \frac{B_{12} + 2B_{66}}{R^2} \frac{\partial^3 w}{\partial x \partial \theta^2} = \rho_T \frac{\partial^2 u}{\partial t^2} \qquad (5.13a)$$

$$\left(\frac{A_{12}+A_{66}}{R}+\frac{B_{12}+B_{66}}{R^2}\right)\frac{\partial^2 u}{\partial x\partial\theta}+\left(A_{66}+\frac{3B_{66}}{R}+\frac{3D_{66}}{R^2}\right)\frac{\partial^2 v}{\partial x^2}$$

$$+\left(\frac{A_{22}}{R^2}+\frac{2B_{22}}{R^3}+\frac{D_{22}}{R^4}\right)\frac{\partial^2 v}{\partial\theta^2}-\left(\frac{B_{12}+2B_{66}}{R}+\frac{D_{12}+2D_{66}}{R^2}\right.$$

$$\left.\frac{\partial^3 w}{\partial x^2\partial\theta}+\left(\frac{A_{22}}{R^2}+\frac{B_{22}}{R^3}\right)\frac{\partial w}{\partial\theta}-\left(\frac{B_{22}}{R^3}+\frac{D_2}{R^4}\right)\frac{\partial^3 w}{\partial\theta^3}=\rho_T\frac{\partial^2 v}{\partial t^2}\right)$$

$$(5.13b)$$

$$B_{11}\frac{\partial^3 u}{\partial x^3}-\frac{A_{12}}{R}\frac{\partial u}{\partial x}+\frac{B_{12}+2B_{66}}{R^2}\frac{\partial^3 u}{\partial x\partial\theta^2}+\left(\frac{B_{12}+2B_{66}}{R}\right.$$

$$+\frac{D_{12}+4D_{66}}{R^2}\left.\right)\frac{\partial^3 v}{\partial x^2\partial\theta}+\left(\frac{B_{22}}{R^3}+\frac{D_{22}}{R^4}\right)\frac{\partial^3 v}{\partial\theta^3}-\left(\frac{A_{22}}{R^2}+\frac{B_{22}}{R^3}\right)\frac{\partial v}{\partial\psi}$$

$$-D_{11}\frac{\partial^4 w}{\partial x^4}-\frac{2(D_{12}+2D_{66})}{R^2}\frac{\partial^4 w}{\partial x^2\partial\theta^2}-\frac{D_{22}}{R^4}\frac{\partial^4 w}{\partial\theta^4}+\frac{2B_{12}}{R}\frac{\partial^2 w}{\partial x^2}$$

$$+\frac{2B_{22}}{R^3}\frac{\partial^2 w}{\partial\theta^2}-\frac{A_{22}}{R^2}w=\rho_T\frac{\partial^2 w}{\partial t^2}\qquad(5.13c)$$

## 5.5 Modal Displacement Forms

The present cylindrical shell is analyzed by applying the Galerkin technique for their vibrations. Before applying this technique, the classical method of separation of variables for PDEs is engaged to split spacial and temporal variables is applied. This process gives birth a set of three ODEs. Modal displacement forms are selected such that they separate the spacial along temporal variables designated by $x, \varphi$ and $t$ respectively. When these forms are substituted for the unknown deformation functions into the shell motion equations, three ODEs are produced in the three dependent variables.

$$u(x, \theta, t) = p_m \frac{d\varphi}{dx} \sin n\theta \cos \omega t$$

$$v(x, \theta, t) = q_m \varphi(x) \cos n\theta \cos \omega t \qquad (5.14)$$

$$v(x, \theta, t) = r_m \varphi(x) \cos n\theta \cos \omega t$$

The amplitude parameters are $p_m, q_m$ and $r_m$. The circumferential and half- axial wave number is donted by $n$ and $m$ respectively. The unknown axial function $[\varphi(x)]$ is used to satify the boundarty conditions. The natural circular frequency for a cylindrical structure is symbolized by $\omega$.

## 5.6 Use of the Galerkin Method

The Galerkin method is applied to obtain the frequency equation in eigen form. The following equations are cropped up after substituting the set of eqn (5.14) into Eqn. (5.13a)–(5.14c).

$$
\left[ A_{11} \frac{d^3\varphi}{dx^3} - n^2 \frac{A_{66}}{R^2} \frac{d\varphi}{dx} \right] p_m - n \left[ \frac{A_{12} + A_{66}}{R} + \frac{B_{12} + 2B_{66}}{R^2} \right] \frac{d\varphi}{dx} q_m
$$

$$
+ \left[ \frac{A_{12}}{R} \left( \frac{d\varphi}{dx} + \varphi(x) \right) - B_{11} \left( \frac{d^3\varphi}{dx^3} + 3 \frac{d^2\varphi}{dx^2} \right) \right. \tag{5.15a}
$$

$$
\left. + n^2 \frac{B_{12} + 2B_{66}}{R^2} \left( \frac{d\varphi}{dx} + \varphi(x) \right) \right] r_{\mathrm{rm}} = -\omega^2 \rho_T p_m \frac{d\varphi}{dx}
$$

$$
n \left( \frac{A_{12} + A_{66}}{R} + \frac{B_{12} + 2B_{66}}{R^2} \right) \frac{d^2\phi}{dx^2} p_m + \left[ \left( A_{66} + \frac{3B_{66}}{R} + \frac{4D_{66}}{R^2} \right) \frac{d^2\varphi}{dx^2} \right.
$$

$$
- n^2 \left( \frac{A_{22}}{R^2} + \frac{2B_{22}}{R^3} + \frac{D_p}{R^4} \right) \varphi(x) \right] q_m + \left[ n \left( \frac{A_{22}}{R^2} + \frac{B_2}{R^3} \right) \varphi(x) \right.
$$

$$
+ n^3 \left( \frac{B_{22}}{R^3} + \frac{D_{22}}{R^4} \right) \varphi(x) - n \left( \frac{B_{12} + 2B_{66}}{R} + \frac{D_{12} + 4D_{66}}{R^2} \right)
$$

$$
\left. \left( \frac{d^2\varphi}{dx^2} + 2 \frac{d\varphi}{dx} \right) \right] r_m = -\omega^2 \rho_T \varphi(x) q_m \tag{5.15b}
$$

$$
\left( -\frac{A_{12}}{R} \frac{d^2\varphi}{dx^2} + B_{11} \frac{d^4\varphi}{dx^4} - n^2 \frac{B_{12} + 2B_{66}}{R^2} \frac{d^2\varphi}{dx^2} \right) p_m
$$

$$
+ \left[ n \left( \frac{A_{22}}{R^2} + \frac{B_{22}}{R^3} \right) \varphi(x) + n^3 \left( \frac{B_{22}}{R^3} + \frac{D_{22}}{R^4} \right) \varphi(x) \right.
$$

$$
\left. - n \left( \frac{B_{12} + 2B_{66}}{R} + \frac{D_{12} + 4D_{66}}{R^2} \right) \frac{d^2\varphi}{dx^2} \right] q_m
$$

$$
+ \left[ -\frac{A_{22}}{R^2} \varphi(x) + \frac{2B_{12}}{R} \left( \frac{d^2\varphi}{dx^2} + 2 \frac{d\varphi}{dx} \right) \right. \tag{5.15c}
$$

$$
\left. - 2n^2 \frac{B_{22}}{R^3} \varphi(x) - D_{11} \left( \frac{d^4\varphi}{dx^4} + 4 \frac{d^3\varphi}{dx^3} \right) - \frac{D_{22}}{R^4} \varphi(x) \right] r_m
$$

$$
= -\omega^2 \rho_T \varphi(x) r_m
$$

Next the Eqns. (5.15a), (5.15b) and (5.15c) are multiplied by $\frac{d\varphi}{dx}$, $\varphi(x)$ and $\varphi(x)$ correspondingly. The resulting equations are integrated with respect to $x$ from 0 to $L$, the following equations are got:

**Table 5.1**    Integral terms

$$I_1 = \int\limits_0^L \frac{d^3\varphi}{dx^3}\frac{d\varphi}{dx}dx \qquad I_2 = \int\limits_0^L \frac{d\varphi}{dx}\frac{d\varphi}{dx}dx \qquad I_3 = \int\limits_0^L \frac{d^2\varphi}{dx^2}\varphi(x)dx$$

$$I_4 = \int\limits_0^L \varphi(x)\varphi(x)dx \qquad I_5 = \int\limits_0^L \frac{d^4\varphi}{dx^4}\varphi(x)dx \qquad I_6 = \int\limits_0^L \frac{d\varphi}{dx}\frac{d\varphi}{dx}dx$$

$$I_7 = \int\limits_0^L \varphi(x)\frac{d\varphi}{dx}dx \qquad I_8 = \int\limits_0^L \frac{d^3\varphi}{dx^3}\frac{d\varphi}{dx}dx \qquad I_9 = \int\limits_0^L \frac{d^2\varphi}{dx^2}\frac{d\varphi}{dx}dx$$

$$I_{10} = \int\limits_0^L \varphi(x)\varphi(x)dx \qquad I_{11} = \int\limits_0^L \frac{d^2\varphi}{dx^2}\varphi(x)dx \qquad I_{15} = \int\limits_0^L \frac{d\varphi}{dx}\varphi(x)dx$$

$$I_{13} = \int\limits_0^L \varphi^2(x)dx \qquad I_{14} = \int\limits_0^L \frac{d^2\varphi}{dx^2}\varphi(x)dx \qquad I_{18} = \int\limits_0^L \frac{d^2\varphi}{dx^2}\varphi(x)dx$$

$$I_{16} = \int\limits_0^L \frac{d^4\varphi}{dx^4}\varphi(x)dx \qquad I_{17} = \int\limits_0^L \frac{d^3\varphi}{dx^3}\varphi(x)dx$$

$$I_{19} = \int\limits_0^L \frac{d^4\varphi}{dx^4}\varphi(x)dx$$

$$\left( A_{11}I_1 - n^2\frac{A_{66}}{R^2}I_2 \right)p_m - n\left( \frac{A_{12}+A_{66}}{R} + \frac{B_{12}+2B_{66}}{R^2} \right)I_2 q_m$$

$$\left[ \frac{A_{12}}{R}(I_6+I_7) - B_{11}(I_8+3I_9) + n^2\frac{B_{12}+2B_{66}}{R^2}(I_6+I_7) \right]$$

$$r_m = -\omega^2\rho_T I_2 p_m \tag{5.16a}$$

$$n\left( \frac{A_{12}+A_{66}}{R} + \frac{B_{12}+2B_{66}}{R^2} \right)I_3 p_m + \left[ \left( A_{66} + \frac{3B_{66}}{R} + \frac{4D_{66}}{R^2} \right)I_3 \right.$$

$$\left. -n^2\left( \frac{A_{22}}{R^2} + \frac{2B_{22}}{R^3} + \frac{D_{22}}{R^4} \right)I_4 \right]q_m \left[ n\left( \frac{A_{22}}{R^2} + \frac{B_{22}}{R^3} \right)I_{10} \right.$$

$$+n^3\left( \frac{B_{m2}}{R^3} + \frac{D_{22}}{R^4} \right)I_{10} - n\left( \frac{B_{12}+2B_{65}}{R} + \frac{D_{12}+4D_{66}}{R^2} \right)$$

$$\left. (I_{11}+2I_{12}) \right]r_m = -\omega^2\rho_T I_4 q_m \tag{5.16b}$$

$$\left( -\frac{A_{12}}{R}I_{18} + B_{11}I_{19} - n^2\frac{B_{12}+2B_{66}}{R^2}I_{18} \right)p_m + \left[ n\left( \frac{A_{22}}{R^2} + \frac{B_{22}}{R^3} \right) \right.$$

$$\left. I_{10} + n^3\left( \frac{B_{22}}{R^3} + \frac{D_{22}}{R^4} \right)I_{10} - n\left( \frac{B_{12}+2B_{66}}{R} + \frac{D_{12}+4D_{66}}{R^2} \right)I_{18} \right]$$

$$q_m + \left[ -\frac{A_{22}}{R^2}I_3 + \frac{2B_{12}}{R}\left(I_{14} + 2I_{15}\right) - 2n^2\frac{D_m}{R^3}I_{13} - D_{11}\left(I_{16} + 4I_{17}\right) \right.$$

$$\left. +2n^2\frac{D_{12} + 2D_{66}}{R^2}\left(I_{14} + 2I_{15}\right) - \frac{D_{22}}{R^4}I_{13} \right]r_m = -\omega^2\rho_T I_{13}r_m$$

$$(5.16c)$$

where the integral terms are listed in Table 5.1

## 5.7 Parametric Study, Validation, and Discussion of Results

The obtained results, in this section are presented through SST. The influence of Poisson's ratios on the natural frequencies of armchair, zigzag and chiral SWCNTs is discussed. The frequency response is investigated for different indices of chiral, zigzag and armchair SWCNTs $(5, 3)$, $(8, 2)$, $(11, 7)$, and $(7, 0)$, $(11, 0)$, $(15, 0)$, and $(6, 6)$, $(8, 8)$, $(10, 10)$, respectively. Before validating the results with published literature, it is necessary to select the parameters. The parameters are taken from the published study of Demir et al. [16]. In the study, the material and geometric parameters have been used $\rho = 2300 \text{ kg/m}^3$, $E = 1012 \text{ N/m}^2$, $L = 10^{-8} \text{ m}$, $d = 33 \times 10^{-9} \text{ m}$, $h = 0.34 \times 10^{-9} \text{ m}$

The schema for armchairs with indices $(5, 5)$, $(7, 7)$, and $(9, 9)$ is demonstrated in Figure 5.3.

In this section, to validate the numerical results of the present model, a comparative study is performed. The results of armchair $(8, 8)$, $(10,10)$, $(12, 12)$, $(14,14)$, $(16,16)$, $(18,18)$, $(20, 20)$, zigzag $(14, 0)$, $(17, 0)$, $(21, 0)$, $(24, 0)$, $(28, 0)$, $(31, 0)$, $(35, 0)$, and chiral $(12, 6)$, $(14, 6)$, $(16, 8)$, $(18, 9)$, $(20, 12)$, $(24, 11)$, $(30, 8)$ are compared with Su and Cho (2021) as shown in Table 5.1. Considering the clamped-free boundary condition, the fundamental frequencies are estimated. The aspect ratio $(L/d = 10)$ is fixed. It is found that the frequencies of three types armchair, zigzag, and chiral SWCNTs are very similar due to the fixed length and clamped-free boundary condition. The present results are closed for three types of SWCNTs and have very little difference. The present frequencies are lower than the corresponding frequencies of Su and Cho (2021). From this table, it is concluded that the two computational methods are well matched. So the selection of comparison becomes very good to apply the present model for particular vibrational frequencies of SWCNTs.

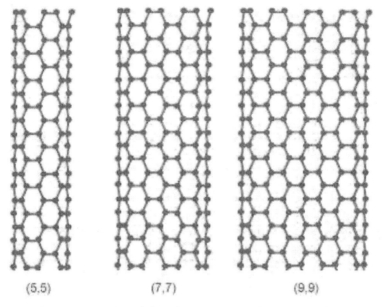

(5.5)                    (7,7)                         (9,9)

**Figure 5.3**    The schematic variation of (5, 5), (7, 7), and (9, 9) armchair nanotubes.

**Table 5.2**    Armchair, zigzag, and chiral (C–F) SWCNT frequency comparison in THz for aspect ratio $L/d = 10$ with Su and Cho [27].

| Clamped-free frequencies (THz), $L/d = 10$ | | | | | | | |
|---|---|---|---|---|---|---|---|
| Armchair SWCNTs | (8, 8) | (10, 10) | (12, 12) | (14, 14) | (16, 16) | (18, 18) | (20, 20) |
| Su and Cho [27] | 3.4886 | 3.4835 | 3.4796 | 3.4767 | 3.4744 | 3.4727 | 3.4712 |
| Present | 3.4830 | 3.4820 | 3.4719 | 3.4728 | 3.4711 | 3.4713 | 3.4704 |
| Zigzag SWCNTs | (14, 0) | (17, 0) | (21, 0) | (24, 0) | (28, 0) | (31, 0) | (35, 0) |
| Su and Cho [27] | 3.4884 | 3.4839 | 3.4794 | 3.4769 | 3.4743 | 3.4727 | 3.4711 |
| Present | 3.4867 | 3.4839 | 3.4736 | 3.4721 | 3.4715 | 3.4719 | 3.4708 |
| Chiral SWCNTs | (12, 6) | (14, 6) | (16, 8) | (18, 9) | (20, 12) | (24,11) | (30, 8) |
| Su and Cho [27] | 3.4894 | 3.4829 | 3.4793 | 3.477 | 3.4743 | 3.4727 | 3.4712 |
| Present | 3.4865 | 3.4813 | 3.4765 | 3.4749 | 3.4732 | 3.4718 | 3.4709 |

## 5.7.1  Influence of Poisson's ratio on the Vibration of Armchair SWCNTs

The natural frequencies of armchair $(6, 6), (8, 8), (10, 10)$, single walled carbon nanotubes versus Poisson's ratio for simply supported-simply supported and clamped-simply supported boundary conditions as sketched in Figures 5.4–5.6. Liang et al. [22] investigated the natural frequencies versus

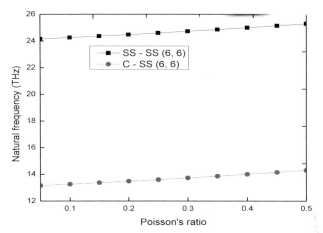

**Figure 5.4** Change of (SS-SS, C-SS) natural frequency with ratios of Poisson for armchair (6, 6) SWCNTs.

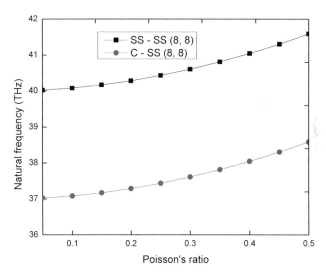

**Figure 5.5** Change of (SS-SS, C-SS) natural frequency with ratios of Poisson for armchair (8, 8) SWCNTs.

Poisson effect using Euler-Bernoulli beam model within the context of a simplified SGT with higher-order inertia. In addition for the prediction of size-dependent natural frequencies, Poisson effect is considered. In Figure 5.6, the frequency value for SS-SS armchair [(6,6),(8,8), (10,10)] at Poisson's ratio (v=0.05,0.1,0.15,0.2,0.25,0.3,0.35,0.4,0.45,0.5) are (24.16, 24.259, 24.365,

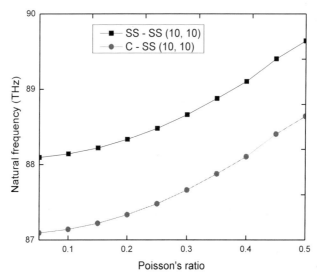

**Figure 5.6**    Change of (SS–SS, C–SS) natural frequency with ratios of Poisson for armchair (10, 10) SWCNTs.

24.478,24.598,  24.725,  24.858,24.998,25.145,25.298), (40.022, 40.081, 40.17, 40.287, 40.434, 40.609, 40.813, 41.044, 41.303, 41.59), (88.094, 88.141, 88.22, 88.334, 88.481, 88.662, 88.877, 89.103, 89.405, 89.642), respectively. Similarly for C-SS armchair [(6, 6),(8, 8), (10, 10)] are (13.16,13.259,13.365,13.478,13.598,13.725,13.858,13.998, 14.145, 14.298), (37.022, 37.081, 37.171, 37.287, 37.434,37.609, 37.813, 38.044,38.303, 38.59), (87.094, 87.141, 87.22, 87.334, 87.481, 87.662, 87.877, 88.103, 88.405, 88.642), respectively. The frequencies of SS-SS (6,6), C-SS (6,6) are in increasing slowly. The variation in Poisson's ratio ($v = 0.05 \sim 0.1$) are as 24.16,24.259. We can observe that there is minute increment is seen and this small increment is captured through ($v = 0.15 \sim 0.5$). It is worth mentioning that the SS-SS frequencies are two times greater than C-SS frequencies. It can be well mentioned that the frequencies increasing on increasing the Poisson's ratio. In fact, inclusion of Poisson's effect during vibration, the effective stiffness increases which increase the natural frequencies.

Now for the case of SS-SS $(8, 8)$, C-SS $(8, 8)$, the frequencies for Poisson's ratio ($v = 0.05, 0.1, 0.15, 0.2, 0.25, 0.3, 0.35, 0.4, 0.45, 0.5$) grows rapidly as compared to SS-SS $(6, 6)$, C $-$ SS$(6, 6)$. The frequencies gap between SS-SS and C-SS is smaller than that of SS-SS $(6, 6)$, C-SS $(6, 6)$. But the frequencies of are higher than SS-SS $(6, 6)$, C-SS $(6, 6)$.

Figure 5.6 displays the frequencies of SS-SS, C-SS $(10, 10)$ versus the effect of Poisson's. Similarly as above like SS-SS $(6, 6)$, C-SS $(6, 6)$ and SS-SS $(8, 8)$, C-SS $(8, 8)$, the frequencies increases on increasing the Poisson's ratio. In this case, $v = 0.45$, the frequency interrupt little bit upward and after that it continues as usual. As seen in Figures 5.4–5.6, the frequency value of armchair tube indicating that the addition of Poisson effect increases the effective stiffness of single walled carbon nanotubes. It is indicated that the frequencies decreases on decreasing the Poisson's ratio from ( $v = 0.5 \sim 0.05$). Also the frequencies of armchair $(10, 10)$ are greater than armchair $(6, 6)$ and $(8, 8)$.

## 5.7.2 Influence of Poisson's ratio on the vibration of zigzag SWCNTs

Figures 5.7–5.9 shows the variation of the frequencies of SS-SS and C-SS zigzag $(7, 0), (11, 0), (15, 0)$ against Poisson's ratio ($v = 0.05, 0.1, 0.15, 0.2, 0.25, 0.3, 0.35, 0.4, 0.45, 0.5$). For initial two values, the numerical values of frequency at $v = 0.005(0.1)$ for SS-SS zigzag $(7, 0)$ are 16.455 (16.533), for SS-SS $(11, 0)$ are $33.219(33.281)$ and for SS-SS zigzag $(15, 0)$ are 55.583 (55.635). The fundamental natural frequencies for $v = 0.005(0.1)$ for C - SS zigzag $(7, 0)$ are 15.455 (15.533), for C - SS $(11, 0)$ are $32.219(32.281)$ and for C - SS zigzag $(15, 0)$ are $54.583(54.635)$. The obtained result for natural

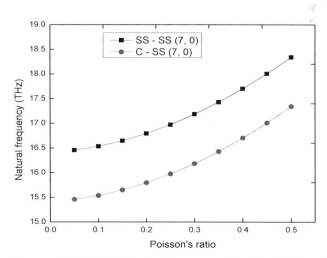

**Figure 5.7** Change of (SS–SS, C–SS) natural frequency with ratios of Poisson for zigzag (7, 0) SWCNTs.

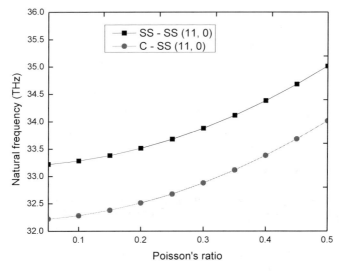

**Figure 5.8**    Change of (SS–SS, C–SS) natural frequency with ratios of Poisson for zigzag (11, 0) SWCNTs.

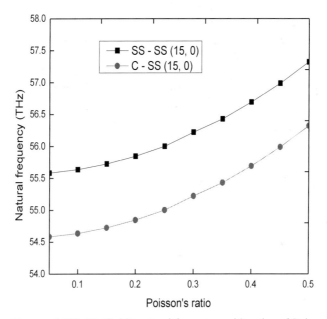

**Figure 5.9**    Change of (SS–SS, C–SS) natural frequency with ratios of Poisson for zigzag (15, 0) SWCNTs.

frequencies portrays a different behavior with Poisson's ratios. The decrease in ratios of Poisson, the frequency increases. The frequency increase slowly upward and also frequencies increases by varying the zigzag index $(7, 0)$ to $(11, 0)$ and $(15, 0)$. We can see that Poisson's ratio as $(v < 0, v = 0, v > 0)$, but the study is restricted for $v > 0$ [25]. If Poisson's ratio as $(v < 0)$, then decrease in the Poisson's ratio gives a decrease in the longitudinal strain and an increase in the lateral strain. For Poisson's ratio $v = 0$, there exist no crosssectional deformation and $v > 0$, then decrease in the Poisson's ratio gives a decrease in both longitudinal strain and lateral strain. This dependency shows that the Poisson's ratio have significant effect on the vibration of single walled carbon nanotubes.

As evidenced by this figure, in practice of mid value the frequencies of $v = 0.25(0.3)$ for SS-SS zigzag $(7, 0)$ are $16.971(17.183)$, for SS-SS $(12, 0)$ are $33.678$ ($33.879$), and for SS-SS zigzag $(15, 0)$ are $56.001(56.221)$

The trend of frequencies for $v = 0.25(0.3)$, for C - SS zigzag $(7, 0)$ are $15.971$ ($16.183$), for C - SS $(11, 0)$ are $32.678(32.879)$, and for C - SS zigzag $(15, 0)$ are $55.001$ ($55.221$). Now for higher Poisson's ratio, $v = 0.25(0.3)$, the frequencies increases fastly. Similarly, the frequency curve of C - SS increase as for C-C case and remaining pattem is same. It is noted that C-SS frequencies is less than C-C frequencies. Moreover, the frequency outcomes for $v = 0.45(0.5)$ for SS-SS zigzag $(7, 0)$ are $18.005(18.338)$, for SS-SS $(11, 0)$ are $34.683(35.015)$ and for SS-SS zigzag $(15, 0)$ are $56.989$ ($57.321$). The frequency value at $v = 0.45(0.5)$, for C - SS zigzag $(7, 0)$ are $17.005$ ($17.338$), for C - SS $(11, 0)$ are $33.683(34.015)$ and for C $-$ SS zigzag $(15, 0)$ are $55.989(56.321)$. For two last ratios, $v = 0.45(0.5)$, the frequencies at the top and in the increasing mode. It is indicated that the [SS-SS, C-SS (10, 10)] is sandwich with [SS-SS, C-SS (7, 0)] and [SS-SS, C-SS (15, 0)].

### 5.7.3 Influence of Poisson's ratio on the vibration of chiral SWCNTs

Here the chiral $(5, 3), (8, 2), (11, 7)$ frequency variation with against Poisson's ratio $(v = 0.05, 0.1, 0.15, 0.2, 0.25, 0.3, 0.35, 0.4, 0.45, 0.5)$ as discussed in Figures 5.10–5.12. The data is sketched in Figures 5.10–5.12 with SS-SS and C-SS boundary condition. Poisson's ratio by virtue of its relation to the moduli, increases with frequency. The Poisson's ratio is a viscoelastic property that depends on temperature, time, strain and strain rate. The variation of the frequencies of SS-SS $[(5, 3), v = 0.15, 0.2]$ are $46.744, 46.837,$

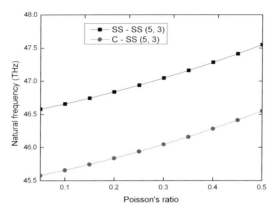

**Figure 5.10**  Change of (SS–SS, C–SS) natural frequency with ratios of Poisson for chiral (5, 3) SWCNTs.

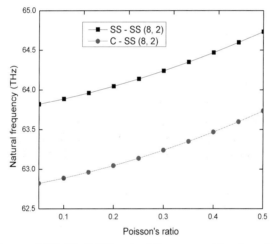

**Figure 5.11**  Change of (SS–SS, C–SS) natural frequency with ratios of Poisson for chiral (8, 2) SWCNTs.

SS-SS $[(8,2), v = 0.15, 0.2]$ are $63.961, 64.045$, SS-SS $[(11,7), v = 0.15, 0.2]$ are $65.941, 66.03$. The results of natural frequencies of C-SS $[(5,3), v = 0.15, 0.2]$ are $45.744, 45.837$, C-SS $[(8,2), v = 0.15, 0.2]$ are $62.961, 63.045$, $C - SS[(11,7), v = 0.15, 0.2]$ are $64.941, 65.03$. As for the case of SS-SS, C-SS $(5,3), (8,2), (11,7)$, the decrease in Poisson's ratios frequencies decreases. It is demonstrated that increasing behavior is seen for different index of chiral single walled carbon nanotubes. Poisson's ratio directly measures the deformation in the material. A high Poisson's ratio

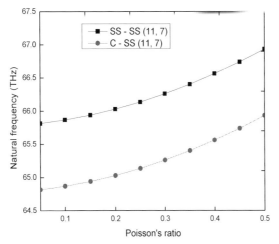

**Figure 5.12** Change of (SS–SS, C–SS) natural frequency with ratios of Poisson for chiral (11, 7) SWCNTs.

denotes that the material exhibits large elastic deformation. Due to these deformation frequencies of carbon nanotubes increases. When the Poisson's ration increases the moduli are expected to become so high and in result sufficiently high frequencies observed. Poisson's ratio is a useful measure of how much a material deforms under stress (stretching or compression). It is important for mechanical engineering as it allows materials to be chosen that suit the desired function.

But once again, the frequencies variation with SS-SS $[(5,3), v = 0.35, 0.4]$ are $47.161, 47.284$, SSSS $[(8, 2), v = 0.35, 0.4]$ are $64.349, 64.468$, SS-SS $[(11, 7), v = 0.35, 0.4]$ are $66.404, 66.504$. The investigated values of chiral frequencies C-SS $[(5, 3), v = 0.35, 0.4]$ are $46.161, 46.284$, C-SS $[(8, 2), v = 0.35, 0.4]$ are $63.349, 63.468, 66.564$, $C - SS[(11, 7), v = 0.35, 0.4]$ are $65.404, 65.564$. The Poisson's ratio was observed to increase the frequency. The estimated frequency values of C-SS are high as compared of SS-SS. The frequency value increases with the increase of indices of single walled carbon nanotubes. The symmetry in the frequencies is seen for all indices $(5, 3), (8, 2), (11, 7)$. The frequencies of chiral $(11, 7)$ is highest that chiral $(5, 3)$ and $(11, 7)$. The frequency pattern with all boundary conditions are seems to be parallel for overall values of Poisson's ratio $(v = 0.05, 0.1, 0.15, 0.2, 0.25, 0.3, 0.35, 0.4, 0.45, 0.5)$.

## 5.8 Conclusions

A numerical approach is developed for the vibration of SWCNTs based on Sander's shell theory. The GLT is used to extract the frequencies of CNTs in the form of eigen value. Some examples are presented in tabular form to compare the results. The accuracy is found for excellent convergence behavior. A detailed parametric study is displayed for the influence of Poisson's ratio on armchair, zigzag, and chiral tubes with simply supported and clamped simply supported edge condition. It is seen that the frequencies increase on increasing the Poisson's ratio. When the inclusion of Poisson's effect is considered during vibration, then the effective stiffness increases which increase the natural frequencies. The estimated frequency values of C–SS are high compared to SS–SS. The frequency value increases with the increase of indices of single-walled carbon nanotubes. The frequency pattern with all boundary conditions seems to be parallel for overall values of Poisson's ratio. In the materials, Poisson's ratio directly measures the deformation and the material shows a large elastic deformation due to high Poisson's ratio. When the Poisson's ratio increases, the moduli are expected to become so high and as a result sufficiently high frequencies are observed. In stretching and compression, Poisson's ratio is a useful measure of how much a material is deformed. It is important for mechanical engineering as it allows materials to be chosen that suit the desired function.

## References

[1] Iijima, S. (1991). Helical microtubules of graphitic carbon. Nature, 354 (7), 56–58

[2] Kuzuo, R., Terauchi, M. T. M., & Tanaka, M. T. M. (1992). Electron energy-loss spectra of carbon nanotubes. *Japanese journal of applied physics*, *31*(10B), L1484.

[3] Tsang, S. C., Harris, P. J. F., & Green, M. L. H. (1993). Thinning and opening of carbon nanotubes by oxidation using carbon dioxide. *Nature*, *362*(6420), 520–522.

[4] Stephan, O., Ajayan, P. M., Colliex, C., Redlich, P., Lambert, J. M., Bernier, P., & Lefin, P. (1994). Doping graphitic and carbon nanotube structures with boron and nitrogen. *Science*, *266*(5191), 1683-1685.

[5] De Heer, W. A., Chatelain, A., & Ugarte, D. (1995). A carbon nanotube field-emission electron source. *science*, *270*(5239), 1179–1180.

[6] Banerjee, J., & F. Williams. (1996). Exact dynamic stiffness matrix for composite Timoshenko beams with applications. Journal of sound and vibration,194(4), 573–585.

[7] Rao, A. M., Richter, E., Bandow, S., Chase, B., Eklund, P. C., Williams, K. A., ... & Dresselhaus, M. S. (1997). Diameter-selective Raman scattering from vibrational modes in carbon nanotubes. *Science*, *275*(5297), 187–191.

[8] Kwon and Tomanek (1998). Electronic and structural properties of multiwall carbon nanotubes. Phys Rev B, 58, 16001–16004.

[9] Leung, A. Y. T., & Kuang, J. L. (2005). Nanomechanics of a multi-walled carbon nanotube via Flugge's theory of a composite cylindrical lattice shell. Physical Review B, 71(16), 165415.

[10] He, X. Q., Eisenberger, M., & Liew, K. M. (2006). The effect of van der Waals interaction modeling on the vibration characteristics of multi-walled carbon nanotubes. Journal of Applied Physics, 100(12), 124317.

[11] Rabczuk, T., Areias, P. M. A., & Belytschko, T. (2007). A meshfree thin shell method for non-linear dynamic fracture. International Journal for Numerical Methods in Engineering, 72(5), 524–548.

[12] Sun, C. Q., & Liu, K. X., (2009).Vibration of multi-walled carbon nanotubes with initial axial force and radial pressure, Journal of Physics D: Applied Physics, 42(17), 175412.

[13] Liew, K. M., & Wang, Q. (2007). Analysis of wave propagation in carbon nanotubes via elastic shell theories. International Journal of Engineering Science, 45(2-8), 227–241.

[14] Hu, Z. L., Guo, X. M., & Ru, C. Q. (2008). Enhanced critical pressure for buckling of carbon nanotubes due to an inserted linear carbon chain. *Nanotechnology*, *19*(30), 305703.

[15] Natsuki, T., Leib, X. W., Ni, Q. Q., & Endo, M. (2010). Vibrational analysis of double-walled carbon nanotubes with inner and outer nanotubes of different lengths. Phys. Lett. A. 374, 4684–4689.

[16] Demir, Ç., Civalek, Ö., & Akgöz, B. (2010). Free vibration analysis of carbon nanotubes based on shear deformable beam theory by discrete singular convolution technique. *Mathematical and Computational applications*, *15*(1) 57–65.

[17] Yang, J., Ke, L. L., & Kitipornchai, S. (2010). Nonlinear free vibration of single-walled carbon nanotubes using nonlocal Timoshenko beam theory. Physica E: Low-dimensional Systems and Nanostructures, 42(5), 1727–1735.

[18] Ghavanloo, E., & Fazelzadeh, S. A. (2012). Vibration characteristics of single-walled carbon nanotubes based on an anisotropic elastic shell model including chirality effect. Applied Mathematical Modelling, 36(10), 4988–5000.

[19] Fang, B., Zhen, Y. X., Zhang, C. P., & Tang, Y. (2013). Nonlinear vibration analysis of double-walled carbon nanotubes based on nonlocal elasticity theory. Applied Mathematical Modelling, 37(3), 1096–1107.

[20] Ansari, R., & Arash, B. (2013). Nonlocal Flügge shell model for vibrations of double-walled carbon nanotubes with different boundary conditions. Journal of Applied Mechanics, 80(2), 021006.

[21] Chawis, T., Somchai, C., & Li, T. (2013). Nonlocal theory for free vibration of single-walled carbon nanotubes, Advanced Materials Research, 747, 257–260.

[22] Liang, X., Hu, S., & Shen, S. (2014). A new Bernoulli–Euler beam model based on a simplified strain gradient elasticity theory and its applications. *Composite Structures, 111*, 317–323.

[23] Ansari, R., & Rouhi, H. (2015). Nonlocal Flügge shell model for the axial buckling of single-walled Carbon nanotubes: An analytical approach. International Journal of Nano Dimension, 6(5), 453–462.

[24] Torkaman-Asadi, M. A., Rahmanian, M., & Firouz-Abadi, R. D. (2015). Free vibrations and stability of high-speed rotating carbon nanotubes partially resting on Winkler foundations. Composite Structures, 126, 52–61.

[25] Faroughi, S., & Shaat, M. (2018). Poisson's ratio effects on the mechanics of auxetic nanobeams. *European Journal of Mechanics-A/Solids, 70*, 8-14.

[26] Eltaher, M. A., Almalki, T. A., Ahmed, K. I., & Almitani, K. H. (2019). Characterization and behaviors of single walled carbon nanotube by equivalent-continuum mechanics approach. Advances in nano research, 7(1), 39.

[27] Su, Y. C., & Cho, T. Y. (2021). Free vibration of a single-walled carbon nanotube based on the nonlocal Timoshenko beam model. Journal of Mechanics, 37, 616–635.

[28] Selim, M. M., Althobaiti, S., Yahia, I. S., Mohammed, I. M., Hussin, A. M., & Mohamed, A. B. A. (2022). Impacts of surface irregularity on vibration analysis of single-walled carbon nanotubes based on Donnell thin shell theory. Advances in nano research, 12(5), 483–488.

[29] A.W. Leissa. *Vibration of Shells.* Acoustical Society of America, Columbus, OH, USA, 1993.

# 6

# Wave Propagation in Single-walled Carbon Nanotubes via Euler Beam Theory

## Abstract

The present chapter is dedicated to frequency analysis of carbon nanotubes. The Euler beam theory is utilized to obtain the small size influence on the variation of density response for armchair, zigzag, and chiral single-walled carbon nanotubes. The two different immoveable boundary conditions are applied at the end of these tubes. These boundary conditions seem to play an important role on the vibration of SWCNTs. The measure of natural frequency depends on the composition of the object, its size, structure, weight, and shape. The impact of density on natural frequency is investigated. The natural frequency of the nanotube is the frequency at which the system resonates. The effect of fundamental natural frequencies against density of C–C and C–F armchair (3, 3), (9, 9), (13, 13), zigzag (4, 0), (8, 0), (10, 0), and chiral (4, 2), (7, 3) and (10, 5) single-walled carbon nanotubes is analyzed. Moreover, the order of index of armchair, zigzag, and chiral nanotubes impresses the frequency values. The frequencies are presented in THz throughout the paper. The vibration frequencies are inversely proportional to mass of the zigzag single-walled carbon nanotubes. So due to the increase of density, the resonant frequency decreases. Convergence of the present study is done to view the accuracy. The available results are well matched with the present provided results.

**Keywords:** Carbon nanotube, vibration, density, order of index, nano-material, Euler beam theory.

## 6.1 Introduction

Over the last decades, there is an ideal investigation among engineers and scientists that have attracted excellent properties such as mechanical, electrical, and thermal. This novel exploration ties a connection between nanoscale and macroscopic world. This is the discovery of carbon nanotubes, reported by Iijima [1] in 1991. Carbon nanotubes have many applications for the development of the next generation in microstructure by different techniques [2–5]. The vibration of carbon nanotubes is extensively studied and has many factors of applications in nano-devices which depend on their natural frequency and resonant frequency [6, 7]. Therefore, the dynamic behavior is studied here.

Based on Euler beam theory, Lordi and Yao [8] identified the thermal vibration frequency and Young's modulus of SWCNTs under different conditions and MD simulation was carried out using universal field forces. Calculation of the cylindrical shell model varies slightly from the results that are obtained by MD simulation. Wildöer et al. [9] studied the atomic structure of single-walled carbon nanotubes and multi-walled carbon nanotubes. It is observed that DWCNTs are used to construct the MWCNTs. Hutchison et al. [10] obtained DWCNTs by arc discharge technique. It was revealed that the inner and outer diameters of DWCNTs are in the range of 1.1–4.2 nm, 1.9–5 nm, respectively, with high-resolution electron microscopy. The other form of CNTs which is called MWCNTs is formed as a result of the rolling of more than one layer of graphene. In the literature, there are two main models about the structure of MWCNTs namely: Parchment and Russian model. A set of coaxially single graphite layer is wrapped around itself in Parchment model, while Russian model suggests a structure like scroll for CNTs. The measured distance on closing these layers of graphite is 0.34 nm. Li and Chou [11] established a connection between structural and molecular mechanics to produce deformation in CNTs. They revealed that the Young's moduli of tube monotonically increase with diameter. The results were compared and found a good convergence with theoretical and experimental work.

Plombon et al. [12] studied the kinetic inductance of individual single-walled carbon nanotubes and high-frequency impedance of carbon nanotube bundles and report that the inductance of the bundle is scaled with the number of individual nanotubes. Liew and Wang [13] discussed the wave dispersion for the highest modes of single- and double-walled CNTs. At one time, thin and thick Love and Cooper Naghdi shell theories have been used to study shear and inertia significance. The feasibility and effective use of elastic models have been explained by comparison of outputs. Wang and Zhang [14] examined the bending and torsional stiffness of single-walled CNTs by applying Flügge shell equations. They implemented three-dimensional model of single-walled CNTs in their work with effect of thickness. Hersam [15] defined the behavior of single-walled and double-walled carbon nanotubes by varying its size. It is observed that if the size of CNT is changed, then the whole structure is changed.

Natsuki et al. [16] predicted the vibrational characteristics of DWCNTs filled with fluid and governing equation derived from Flügge's shell equation for the vibration of CNTs. They investigated vibrational modes, influence of parameters, and fluid properties of CNT. Sakhaee-Pour et al. [17] obtained the estimation of continuum modeling and compared these results with MD simulation methods which demonstrate current trends of continuum mechanics model development for investigators and scientists. Murmu and Pradhan [18] reported the nonlocal theory for the smooth conduction of vibration of SWCNTs. Thermal vibrational response of SWCNTs is obtained by utilizing DQM. The effects of Winkler constant, temperature change, and nonlocal small-scale length and vibration modes of CNT are investigated. Ghorbanpour Arani et al. [19] used Donnel shell model to measure the transverse displacements of single- and double-walled carbon nanotubes, the results of which are compared with Euler–Bernoulli and Timoshenko beam models. They observed that beam models predict the lowest frequency and shell models predict the highest. Simsek [20] employed nonlocal Euler–Bernoulli model attributed by end supports to exploit the forced vibrations of single-walled CNTs and fabricated results with aspect ratio and velocity in some detail. Among other things, load velocities and excitation of frequencies are equally important in nano/micro-science. Rafiee and Moghadam [21] employed a 3D-FEM with reinforced polymer CNT. The behavior of reinforced polymer CNT is evaluated both in axial and transverse directions subjected to high strain loading and impact of buckling that are stimulated at micro-scale. Ansari and Rouhi [22] studied the scale effect on resonant frequency of single-walled CNTs. They derived nonlocal shell theory by incorporating elasticity equations into Flügge shell theory. They found that rather than chirality, suitable values of nonlocal parameter have more influence on boundary conditions. Storch and Elishkoff [23] investigated the vibrations of cantilevered DWCNT of a virus and bacterium of the outer nanotube. Here, the exact expressions considered for the characteristic frequency equation and mode shape were extracted by Euler–Bernoulli beam theorem. Alibeigloo and Shaban [24] inquired the significance of nonlocal parameter by employing three-dimensional elastic theories with Fourier expansion on vibrations of CNTs. They concluded that by increasing the value of non-local parameter, the frequency follows a decreasing pattern. Chawis et al. [25] reported a nonlocal theory with scale length to conduct vibration of SWCNTs with Euler beam theory using nonlocal parameter. The results are obtained by classical solutions and compared with the results of FEM. In this study, effects of different geometrical boundary conditions and tube chirality have been considered. They reported that a different variation in the frequency is observed with increasing the length, diameter, and atomic arrangements. Moreover, a new pattern of frequencies is observed with increasing the nonlocal parameter. Recently, CNTs are used on a large scale applying synthesis method with various lengths, diameters, and chiralities. There are many methods for the production of SWCNTs for the distribution of diameter and chiralities which are used in transistors and sensors [26, 27]. Rayleigh and TBM [28] conducted a study using nonlocal theory sound wave propagation in zigzag DWCNTs. They investigated the effects of nonlocal parameter for chirality and aspect ratio on the frequency of DWCNTs. Wu and Lai [29] conducted free vibration of SWCNTs based on the principal of virtual displacement. The strong formulation was made on the principle of Hamiltonian. Rouhi et al. [30] simulated the nonlocal model using Flügge shell theory and to solve the MWCNTs problem with the conjunction of layer combination, the Rayleigh–Ritz method is employed. Azimzadeh and Fatahi-Vajari [31] investigated the coupled axial-radial (CAR) vibration of SWCNTs. The fourth-order PDEs are utilized to govern the coupled vibration of SWNTs with scale parameter. The doublet mechanics is used for the first time for the axial vibration of CNTs. The effect of natural frequencies with scale parameter is seen. Khosravi et al. [32] selected the Eringen's theory for the torsional vibration of SWCNTs. The elastic medium is embedded for SWCNTs under clamped–clamped boundary condition. The difference between torsional and forced vibration with exponential and harmonic load is investigated. Wang and Hu [33] used nano-electromechanical systems (NEMS) with different parameters to check the complex behavior of CNTs. The role of NEMS for parametric excitation and external excitation is seen. Selim [34] obtained the natural frequencies analytically based on Hamilton's principle. The closed-form solution is derived to observe the torsional vibrations of SWCNTs.

Here, the vibrations of single-walled CNTs have been analyzed with clamped–clamped and clamped-free conditions with different chiral indices. The governing equation of Euler beam model has been developed for the vibrations of SWCNTs. Also the frequency characteristics of CNTs for the variation of density versus natural frequency are either not established or assumed very little attention. After presenting the convergence of presented model, the influences of two boundary conditions, density variation, and natural frequency on characteristics of armchair, zigzag, and chiral are investigated in detail.

## 6.2 Theoretical Formulation

Different models are developed for the mechanical behavior (vibration, bending, and buckling) of carbon-based micro- and nanostructures. In this chapter, vibration of single-walled carbon nanotubes based on classical theory named Euler–Bernoulli beam theory is proposed. The index description and orientation of hexagonally graphene sheet for armchair ($n = m$), zigzag ($m = 0$), and chiral ($n \neq m$) are shown in Figure 6.1. The single-walled carbon nanotubes are constructed from the curling of sheet of graphene into hollow cylinder. There are infinitely geometrical shapes that occur due to the rolling up of graphene sheet. The actual discrete SWCNT can be visualized in Figure 6.2.

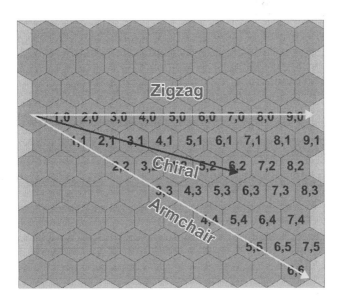

**Figure 6.1**  Description of armchair, zigzag, and chiral indices on graphene sheet of SWCNTs.

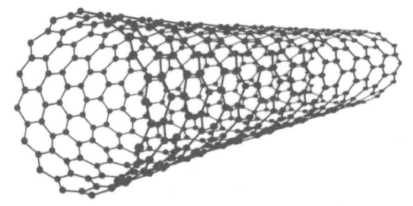

**Figure 6.2**    Artistic representation of SWCNT.

## 6.2.1 Algorithm description

The classical theory states that when the stress is applied on a given point, then the resultant strain is produced at that point. The constitutive equation of the stress and strain relationship in one-dimensional case is as below [35, 36].

$$\sigma_{xx} = E\varepsilon_{xx} \tag{6.1}$$

Where $\sigma_{xx}$ is the axial stress, $\varepsilon_{xx}$ is the axial strain, $E$ the Young modulus. Assume that the displacement of beam along y axis is $u(x,t)$ in terms of spatial coordinate $x$ and time variable $t$. For transversely vibration of carbon nanotubes, the equilibrium conditions of Euler-Bernoulli beam can be written as

$$\frac{\partial M(x,t)}{\partial x^2} = \rho A \frac{\partial^2 u(x,t)}{\partial t^2} \tag{6.2}$$

According to the linear theory of Euler–Bernoulli beam, the strain–displacements and the moment are given by

$$\varepsilon = -y\frac{\partial^2 u(x,t)}{\partial x^2} \tag{6.3}$$

$$M(x,t) = \int_A y\sigma dA \tag{6.3a}$$

Multiplying on both sides of eqn (6.3) by $y$ and integrating over the cross-section area ($A$) of the beam, we obtain

$$\int_A y\sigma_{xx}dA = \int_A Ey\varepsilon dA \tag{6.4}$$

After some mathematical manipulations, we have

$$M(x,t) + EI$$
$$\partial^2 u(x,t)/\partial x^2 = 0 \tag{6.5}$$

By performing the differentiating of this equation with respect to the variable $x$ twice, we have

$$EI\frac{\partial^4 u(x,t)}{\partial x^4} + \rho A\frac{\partial^2 u(x,t)}{\partial t^2} = 0 \tag{6.6}$$

We can write as

$$\rho A\frac{\partial^2 u}{\partial t^2} + \frac{\partial^2}{\partial x^2}\left[EI(x)\frac{\partial^2 u}{\partial x^2}\right] = 0 \tag{6.7}$$

where $\rho$ and $I$ stand for the density and moment of inertia of CNT. Two systems of ordinary differential equations (ODEs) are derived using the Fourier method of variational separation. In this system, two terms are related to the spatial variable $x$ and temporal variable, respectively.

$$u(x,t) = \gamma(x)S(t) \tag{6.8}$$

$$\tau(x)\frac{\partial^2 u}{\partial t^2} + \frac{\partial^2}{\partial x^2}\left[EI(x)\frac{\partial^2 u}{\partial x^2}\right] = 0 \tag{6.9}$$

$$\rho A\frac{\partial^2}{\partial t^2}\gamma(x)S(t) + \frac{\partial^2}{\partial x^2}\left[EI\frac{\partial^2}{\partial x^2}\gamma(x)S(t)\right] = 0 \tag{6.10}$$

$$\rho A\gamma(x)\frac{d^2 S}{dt^2} + EIS(t)\frac{d^4\gamma}{dx^4} = 0 \tag{6.11}$$

$$EIS(t)\frac{d^4\gamma}{dx^4} = -\rho A\gamma(x)\frac{d^2 S}{dt^2} \tag{6.12}$$

For harmonic response

$$S(t) = e^{i\omega t} \text{ or } \cos\omega t \text{ or } \sin\omega t \tag{6.13}$$

Substituting eqn (6.13) into eqn (6.12), the relation can be written as

$$\rho A \gamma \left( -\omega^2 \cos \omega t \right) + EI \cos \omega t \frac{d^4 \gamma}{dx^4} = 0 \tag{6.14}$$

$$\frac{d^4 \gamma}{dx^4} - \frac{\rho A \omega^2}{EI} \gamma(x) = 0 \tag{6.15}$$

$$\frac{d^4 \gamma}{dx^4} - \lambda^4 \gamma(x) = 0 \tag{6.16}$$

Here $\gamma(x)$ denotes the mode shape (Eigen shape)
For parameter $\lambda$

$$\lambda^4 = \frac{\rho A \omega^2}{EI} \tag{6.17}$$

The general solution of fourth-order ODE is postulated as

$$\gamma(x) = q_1 \sin \lambda x + q_2 \cos \lambda x + q_3 \sin \lambda x + q_4 \cosh \lambda x \tag{6.18}$$

where $q_1$, $q_2$, $q_3$, and $q_4$ are the unknown constants.
Eqn (6.16) becomes

$$\gamma^{iv}(x) - \lambda^4 \gamma(x) = 0 \tag{6.19}$$

## 6.2.2 Numerical technique

Wave propagation approach is used to study the vibrational behavior of SWCNTs. Before this work, the current approach was successfully used for vibration and buckling analysis of cylindrical shell and carbon nanotubes.

$$\gamma(x) = e^{-i\Gamma_m} \tag{6.20}$$

For vibrating carbon nanotubes, the axial wavenumber $\Gamma_m$ related to support conditions applied on both sides of SWCNTs and $m$ denotes axial half-wave number.

$$\gamma^{iv}(x) = \Gamma_m^4 e^{-i\Gamma_m x} \tag{6.21}$$

Substituting eqn (6.21) into eqn (6.19), we have

$$\Gamma_m^4 e^{-i\Gamma_m x} - \lambda^4 e^{-i\Gamma_m x} = 0 \tag{6.22}$$

$$\lambda^4 - \Gamma_m^4 \tag{6.23}$$

By using eqn (6.17), we can write as

$$\frac{\rho A \xi \omega^2}{EI} = \Gamma_m^4 \tag{6.24}$$

## 6.2.3 Boundary conditions

Appropriate material properties and boundary conditions are applied and then the model is solved for natural frequencies of SWCNTs of different indices (5, 5), (7, 7), (9, 9) for armchair, for zigzag (8, 0), (15, 0), (20, 0), and for chiral (12, 5), (22, 7), (25, 10) SWCNTs.

From eqn (6.24) For C–C,

$$\frac{\rho A \xi \omega^2}{EI} = \left(\frac{(2n + 1)\pi}{2L}\right)^4 \tag{6.25}$$

Here $\quad \Gamma_n = \dfrac{(2n + 1)\pi}{2L} \quad$ (C-C boundary condition)

For C–F

$$\frac{\rho A \xi \omega^2}{EI} = \left(\frac{(2n - 1)\pi}{2L}\right)^4 \tag{6.26}$$

where $\Gamma_= = \frac{(2n-1)\pi}{2L}$ (C-F boundary condition)

From Eq. (24) the fundamental natural frequencies are calculated where $\omega = 2\pi f$. There exists uncertainty in defining the nanotube thickness. Here, we apply relations from [37].

$$m = \rho A = 2.4 \times 10^{-24} d\,[\,\text{kg/nm}] \tag{6.27}$$

$$EI = 428.48 d^2 - 397.08 d + 109.24 \left[\text{kgnm}^3/\text{s}^2\right] \tag{6.28}$$

Diameter of nanotubes is indicated by $d$ and can be calculated from translation indices $(n, m)$ by relation.

$$d = 2R = a_0 \sqrt{3\left(m^2 + n^2 + nm\right)}\big/\pi \tag{6.29}$$

where the carbon–carbon bond length ($a_0$ = 1.42 Å).

## 6.3 Modeling Results and Discussions

### 6.3.1 Evaluation parameters

On the basis of the result obtained through Euler beam theory, the effect of density variation with clamped–clamped and clamped-free conditions is observed. To compare the fundamental frequencies of SWCNTs of the present solution, let us define the parameters [29]. The assumed parameters are as slenderness ratio $L/d = 10$, mass density $\rho = 2300$ kg/m$^3$, and Poisson ratio $\nu = 0.3$.

### 6.3.2 Modeling validation

In order to give an idea about the accuracy to the readers, a comparison is done with the vibration case of Euler beam theory. By using this theory, the natural frequency for SWCNTs is computed. The results are compared with the study of Sakhaee-Pour et al. [17]. They used beam element method to investigate the vibrational behavior of SWCNTs. The results of armchair and zigzag are tested for different modes which are very familiar with the outcomes of Sakhaee-Pour et al. [17]. The convergence of the present theory is checked satisfactory with existing model. The advantages of the proposed model are that it can be further extended for the vibration of three types of SWCNTs with different boundary conditions (Tables 6.1 and 6.2).

**Table 6.1**   Frequency convergence in THz of armchair SWCNTs with Sakhaee-Pour et al. [17].

| Mode no. | Frequencies (THz) | | | |
|---|---|---|---|---|
| | Clamped–clamped | | Clamped-free | |
| | Present | Sakhaee-Pour et al. [17] | Present | Sakhaee-Pour et al. [17] |
| 1 | 0.543 | 0.557 | 0.103 | 0.104 |
| 2 | 0.556 | 0.557 | 0.104 | 0.104 |
| 3 | 1.26 | 1.28 | 0.560 | 0.562 |
| 4 | 1.27 | 1.28 | 0.561 | 0.562 |
| 5 | 1.31 | 1.32 | 0.64 | 0.65 |
| 6 | 1.89 | 1.91 | 0.943 | 0.947 |
| 7 | 2.12 | 2.13 | 1.33 | 1.34 |
| 8 | 2.13 | 2.13 | 1.34 | 1.34 |
| 9 | 2.13 | 2.15 | 1.93 | 1.95 |
| 10 | 2.14 | 2.15 | 2.07 | 2.08 |

**Table 6.2** Frequency convergence in THz of zigzag SWCNTs with Sakhaee-Pour et al. [17].

| Mode no. | Frequencies (THz) | | | |
|---|---|---|---|---|
| | Clamped–clamped | | Clamped-free | |
| | Present | Sakhaee-Pour et al. [17] | Present | Sakhaee-Pour et al. [17] |
| 1 | 0.425 | 0.427 | 0.071 | 0.072 |
| 2 | 0.426 | 0.427 | 0.071 | 0.072 |
| 3 | 1.03 | 1.05 | 0.420 | 0.421 |
| 4 | 1.05 | 1.05 | 0.420 | 0.421 |
| 5 | 1.26 | 1.27 | 0.627 | 0.628 |
| 6 | 1.77 | 1.78 | 0.881 | 0.882 |
| 7 | 1.82 | 1.83 | 1.06 | 1.07 |
| 8 | 1.83 | 1.83 | 1.07 | 1.07 |
| 9 | 2.53 | 2.54 | 1.86 | 1.87 |
| 10 | 2.68 | 2.69 | 1.87 | 1.87 |

### 6.3.3 Effect of fundamental natural frequencies against density on vibration of armchair SWCNTs

Figures 6.3–6.5 show that the effect of fundamental natural frequencies against density of C – C and C-F armchair $(3,3), (9,9), (13,13)$. The

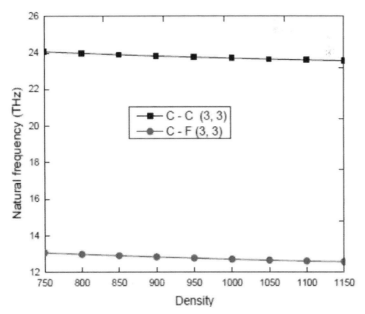

**Figure 6.3** Density influence of armchair (3, 3) SWCNTs on (C–C, C–F) frequencies.

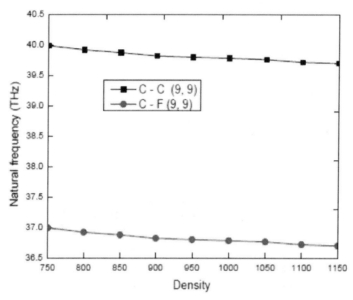

**Figure 6.4**   Density influence of armchair (9, 9) SWCNTs on (C–C, C–F) frequencies.

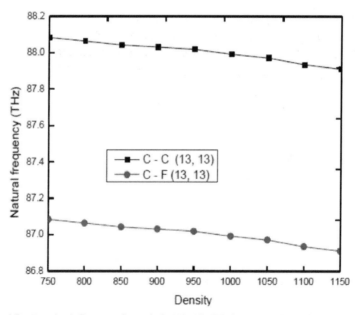

**Figure 6.5**   Density influence of armchair (13, 13) SWCNTs on (C–C, C–F) frequencies.

measure of natural frequency depends on the composition of the object, its size, structure, weight and shape. The impact of density on natural frequency is investigated. This model can predict the frequencies of nanotubes for varying the densities ($\rho = 750 \sim 1150$). The frequencies of armchair C $-$ C$[(3,3),(9,9),(13,13)]$ at ($\rho = 750$) are $24.067, 39.992, 88.083$ and frequencies of armchair C-F $[(3, 3), (9, 9), (13, 13)]$ at initial density ($\rho = 750$) are $13.067, 36.992, 87.083$. It is clearly seen that, the natural frequency curves of armchair nanotube are different over different index for initial same density. Now we see the behavior of frequencies on increasing the densities. The variation in frequencies of armchair C $-$ C$[(3,3),(9,9),(13,13)]$ at ($\rho = 800$) are $[23.982, 39.921, 88.063]$ and of armchair C-F $[(3,3),(9,9),(13,13)]$ at ($\rho = 800$) are $[23.982, 39.921, 88.063]$. It is observed that the frequencies minutely decreases when the density increases from ($\rho = 750 \sim 800$). Therefore, density increase results in resonant frequency decrease. Consequently, natural frequencies also decrease. Furthermore, for tiny values of half wave number, it can be demonstrated. It can be seen that present model provide the frequencies in decreasing manner on increasing the index of armchair tube. Once again, the final frequencies at ($\rho = 1150$)C $-$ C$[(3,3),(9,9),(13,13)]$ are $23.572$, $39.701$, $87.911$ and frequencies of armchair C-F $[(3,3),(9,9),(13,13)]$ are $12.572, 36.701, 86.911$. It is noted that the frequencies decreases very slow down for initial values from ($\rho = 750 \sim 950$) for both C $-$ C and C $-$ F boundary conditions. The frequencies for armchair $(3,3)$, seems as a straight line with slow down for density ($\rho = 750 \sim 1150$) and in case of armchair $(9, 9)$, the frequencies falls down rapidly. The behavior of frequencies is entirely different for armchair $(13, 13)$. Due to large index of armchair $(13, 13)$, the frequency decreases fastly compared to armchair $(3, 3)$ and $(9, 9)$. Furthermore, for tiny values of density, it can be demonstrated that density ($\rho = 750 \sim 900$), effect of initial frequencies $f$(THz)) are weak and seems almost the same values of frequency. However, when the value of $\rho$ is larger than $950 \sim 1150$), the differences become different. The trend of the frequencies is same as the C $-$ C$[= (3, 3), (9, 9), (13, 13)]$, but it is noted that with each index the C-F values is lower than those of corresponding C $-$ C frequencies. It is also observed that these frequencies have a paramount impact on the vibration of CNTs and this is due to the constraints which are applied on the edges of CNT.

### 6.3.4 Effect of Fundamental Natural Frequencies against Density on Vibration of Zigzag SWCNTs

In Figures 6.6–6.8 frequencies shows the impact of vibrating zigzag $(4, 0), (8, 0), (10, 0)$ single walled carbon nanotubes versus variation in density. The calculated results from present model for zigzag index $C-C(= 4, 0)$ at $\rho(= 750, 800, 850, 900, 950, 1000, 1050, 1100, 1150)$, the first nine frequencies are $16.411, 16.301, 16.182, 16.059, 15.918, 15.769, 15.548, 15.307,$ $14.998$ and for zigzag index $C - C(= 8, 0)$, the frequencies are $33.162, 33.047, 32.882, 32.659, 32.418, 32.196, 32.008, 31.787, 31.498$ and frequencies for zigzag index $C - C(= 10, 0)$, are $55.445, 55.301,$ $55.182, 55.059, 54.918, 54.769, 54.548, 54.307, 53.998$. The natural frequency of the nanotube is the frequency at which the system resonates. Here, the natural frequency is gritty by two factors: the amount of mass, and the stiffness of the nanotube, which acts as a spring. The vibration frequencies are inversely proportional to mass of the zigzag single walled carbon nanotubes. So due to increase of density results

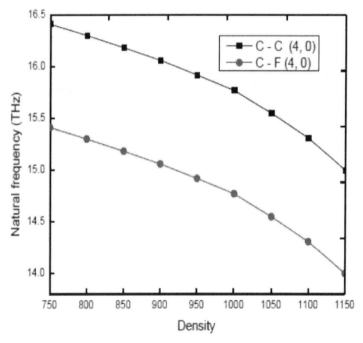

**Figure 6.6**   Density influence of zigzag (4, 0) SWCNTs on (C–C, C–F) frequencies.

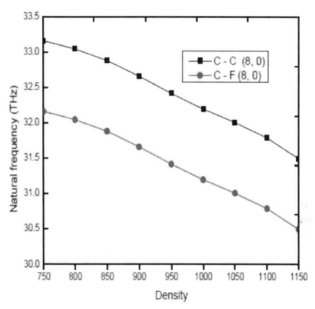

**Figure 6.7**    Density influence of zigzag (8, 0) SWCNTs on (C–C, C–F) frequencies.

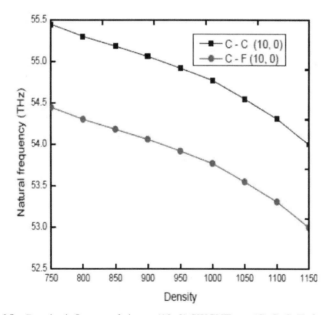

**Figure 6.8**    Density influence of zigzag (10, 0) SWCNTs on (C–C, C–F) frequencies.

the decrease in resonant frequency. Consequently, natural frequencies also decrease. We turn next to examine in more detail zigzag index $C - F(= 4,0)$ at $\rho(= 750 \sim 1150)$ the first none frequencies are $15.411, 15.301, 15.182, 15.059, 14.918, 14.769, 14.548, 14.307, 13.998$, and for zigzag index $C - F(= 8,0)$, the frequencies are $32.162$ , $32.047, 31.882, 31.659, 31.418, 31.196, 31.008, 30.498$ and for zigzag index $C - F(= 10,0)$, are $54.445, 54.301, 54.182, 54.059, 53.918, 53.769, 53.548$, $53.307, 52.998$. Similar behavior of frequencies decreases with increases of density is observed for $C - F$ boundary condition. It can be seen that the frequencies decrease so fastly for C-C, C-F $(4,0), (8,0), (10,0)$ indices of zigzag. The behavior of frequencies decline is same for zigzag indices $(4,0), (8,0), (10,0)$. But the frequencies increase on increase the index from $(4,0)$ to $(8,0)$ and $(10,0)$. The zigzag index $(8,0)$ for two boundary conditions is sandwich between zigzag index $(8,0)$ and $(10,0)$. In addition, it summarized that the frequency gap of zigzag index $(8,0)$ is less than other two index. The frequency of C-F boundary condition is lesser from C-C condition.

It is clearly explicit that density of nanotubes has considerable influence on the natural frequencies of shell. It is noted that natural frequencies increases on increasing the density. With higher density, the frequencies decrease. It is mainly because of the fact that the mass of the system has reverse effect on the natural frequency.

### 6.3.5  Effect of fundamental natural frequencies against density on vibration of chiral SWCNTs

Figures 6.9–6.11 demonstrate the natural frequencies (THz) of chiral single-walled carbon nanotubes based on Euler beam model against the variation in density. The proposed model can accurately predict the acquired results of material data point. The influence of density for natural frequencies is investigated with simply supported boundary condition and clamped-free boundary condition. Moreover, the order of index such as (4, 2), (7, 3), and (10, 5) of the chiral tube impresses the frequency values. The corresponding frequency has been sketched in Figures 6.9–6.11. It is noted that, with the increase of density, fundamental frequencies of the SWCNTs decrease as C–C (C–F) = $(4, 2) f \sim 46.506(45.506)$, C–C (C–F) = $(7, 3) f \sim 63.761(62.761)$, C–C = $(10, 5) f \sim 65.778(64.778)$ at $\rho = 750$ decrease to C–C (C–F) = (4, 2) $f \sim 46.441(45.441)$, C–C (C–F) = (7, 3) $f \sim 63.713(62.713)$, C–C (C–F) = (10, 5) $f \sim 65.76(64.76)$ at $\rho = 800$ and decrease to C–C (C–F) = (4, 2)

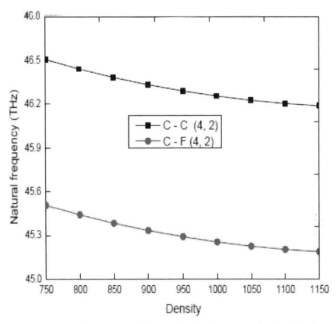

**Figure 6.9** Density influence of chiral (4, 2) SWCNTs on (C–C, C–F) frequencies.

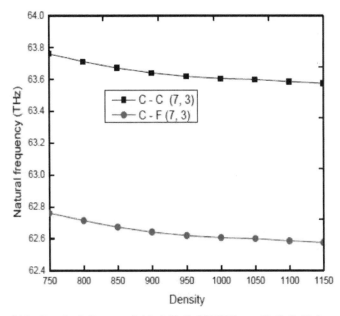

**Figure 6.10** Density influence of chiral (7, 3) SWCNTs on (C–C, C–F) frequencies.

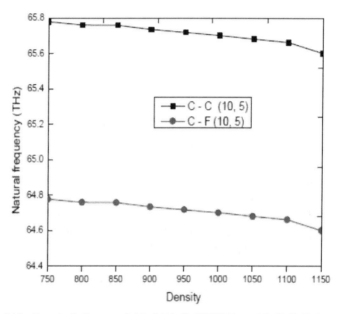

**Figure 6.11**    Density influence of chiral (10, 5) SWCNTs on (C–C, C–F) frequencies.

$f \sim 46.383$ 45.383, C–C (C–F) = (7, 3) $f \sim 63.672$, (62.672), C–C (C–F) = (10, 5) $f \sim 65.759(64.759)$ at $\rho = 850$ for chiral SWCNTs as shown in Figures 6.9–6.11. Also the behavior of frequencies is in decreasing manners for final density as C–C (C–F) = (4, 2) $f \sim 46.188$ (45.188), C–C (C–F) = (7, 3) $f \sim 63.575(62.575)$, C–C (C–F) = (10, 5) $f \sim 65.602$ (64.602) at $\rho = 1100$ for chiral SWCNTs. In these figures, a reasonable decline in curve suggests that frequencies decrease with increase in density $\rho = 750 - 1150$. From our results, one can easily conclude that the decline in frequencies of curves (4, 2), (7, 3), and (10, 5) is as follows: (4, 2) < (7, 3) < (10, 5). Interestingly, in all cases C–F frequency curves are lower than C–C, but follow the same pattern as discussed earlier. The frequency pattern of chiral (4, 2) and (7, 3) are the same.

In comparison of frequencies with armchair, zigzag, and chiral tubes, we get a new phenomenon according to the structure of the tube. Due to an easy deformation in the cross-section of armchair tube, the frequencies are totally different than the zigzag and chiral tubes. The reason is that there is no deformation in the cross-section of zigzag and chiral. Sometimes, when cross-section is deformed but does not remain circular, then different irregular circumferential waveforms, torsional and longitudinal modes can be observed. It is predicted that cross-section has no deformation in zigzag and chiral tubes. Moreover, the cross-sectional deformation of chiral is greater than the zigzag.

## 6.4 Conclusion

The frequency impact of armchair, zigzag, and chiral single-walled carbon nanotubes is explored based on Euler beam model. This continuum model is used to determine the frequencies under clamped–clamped and clamped-free edge conditions. The frequencies are shown in THz throughout the study. The outcomes of results from computer software MATLAB are tested with other computational techniques and found valid which shows that this model can predict the frequencies of nanotubes with varying densities of fundamental natural frequencies against density. Here, the natural frequency is gritty by two factors: the amount of mass and the stiffness of the nanotube, which acts as a spring. Therefore, density increase results in resonant frequency decrease. Consequently, natural frequencies also decrease. Furthermore, for tiny values of half-wave number, it can be demonstrated. It can be seen that the present model provides the frequencies in decreasing manner on increasing the index of armchair tube. It is clearly explicit that density of nanotubes has considerable influence on the natural frequencies of the shell. It is noted that natural frequencies increase on increasing the density. With higher density, the frequencies decrease. It is mainly because of the fact that the mass of the system has reverse effect on the natural frequency, but it is noted that with each index the C–F values are lower than those of corresponding C–C frequencies. It is also observed that these frequencies have a paramount impact on the vibration of CNTs and this is due to the constraints which are applied on the edges of CNT.

## References

[1] Iijima, S. (1991). Helical microtubules of graphitic carbon. Nature, 354 (7), 56–58.

[2] Chernozatonskii, L. A. (1992). Carbon nanotube elbow connections and tori. *Physics Letters A, 170*(1), 37–40.

[3] Endo, M., Takeuchi, K., Igarashi, S., Kobori, K., Shiraishi, M., & Kroto, H. W. (1993). The production and structure of pyrolytic carbon nanotubes (PCNTs). *Journal of Physics and Chemistry of Solids, 54*(12), 1841–1848.

[4] Tersoff, J., & Ruoff, R. S. (1994). Structural properties of a carbon-nanotube crystal. *Physical Review Letters, 73*(5), 676.

[5] Hiura, H., Ebbesen, T. W., & Tanigaki, K. (1995). Opening and purification of carbon nanotubes in high yields. *Advanced Materials, 7*(3), 275–276.

[6] Banerjee, J., & F. Williams. (1996). Exact dynamic stiffness matrix for composite Timoshenko beams with applications. Journal of sound and vibration,194(4), 573– 585.

[7] Rao, A. M., Richter, E., Bandow, S., Chase, B., Eklund, P. C., Williams, K. A., ... & Dresselhaus, M. S. (1997). Diameter-selective Raman scattering from vibrational modes in carbon nanotubes. *Science, 275*(5297), 187–191.

[8] Lordi, V. & Yao. N. (1998). Young's modulus of single-walled carbon nanotubes. J. Appl. Phys., 84, 1939–1943.

[9] Wildöer J. W. G., Venema, L. C., Rinzler, A. G., Smalley, R. E., & Dekker, C. (1998). Electronic structure of atomically resolved carbon nanotubes. Nature, 391, 59–62.

[10] Hutchison, J.L., Kiselev, N.A., Krinichnaya, E.P., Krestinin, A.V., Loutfy, R.O., Morawsky, A.P., Muradyan, V.E., Obraztsova, E. D., Sloan, J., Terekhov, S.V. & Zakharov, D.N., (2001). Double-walled carbon nanotubes fabricated by a hydrogen arc discharge method. Carbon, 39, 761.

[11] Li, C., & Chou, T. W. (2003). A structural mechanics approach for the analysis of carbon nanotubes. International Journal of Solids and Structures, 40(10), 2487–2499.

[12] Plombon, J. J., O'Brien, K. P., Gstrein, F., Dubin, V. M., & Jiao, Y. (2007). High-frequency electrical properties of individual and bundled carbon nanotubes. Applied physics letters, 90(6), 063106.

[13] Liew, K. M., & Wang, Q. (2007). Analysis of wave propagation in carbon nanotubes via elastic shell theories. International Journal of Engineering Science, 45(2-8), 227–241.

[14] Wang, C. Y. & Zhang, L. C. (2007). Modeling the free vibration of single-walled carbon nanotubes. 5th Australasian Congress on Applied Mechanics, ACAM, Brisbane, Australia, 10–12.

[15] Hersam, M. C. (2008). Progress towards monodisperse single-walled carbon nanotubes. *Nature nanotechnology, 3*(7), 387–394.

[16] Natsuki, T., Ni, Q. Q., & Endo, M. (2008). Analysis of the vibration characteristics of double-walled carbon nanotubes. Carbon, 46(12), 1570–1573.

[17] Sakhaee-Pour, A., Ahmadian, M. T., and Vafai, A., 2009, "Vibrational analysis of single-walled carbon nanotubes using beam element," Thin-Walled Structures, 47(6–7), pp. 646–652.

[18] Murmu, T. & S. C. Pradhan, (2009). Thermo-mechanical vibration of a single-walled carbon nanotube embedded in an elastic medium based on nonlocal elasticity theory. Computational Materials Science, 46(4), 854–859

[19] Ghorbanpourarani, A., Mohammadimehr, M., Arefmanesh, A., & Ghasemi, A. (2010). Transverse vibration of short carbon nanotubes using cylindrical shell and beam models. Proceedings of the Institution of Mechanical Engineers, Part C: Journal of Mechanical Engineering Science, 224(3), 745–756.

[20] Simsek, M. (2010). Vibration analysis of a single-walled carbon nanotube under action of a moving harmonic load based on nonlocal elasticity theory. Physica E, 43,182–191.

[21] Rafiee, R., & Moghadam, R. M. (2012). Simulation of impact and post-impact behavior of carbon nanotube reinforced polymer using multi-scale finite element modeling. Computational Materials Science, 63, 261–268.

[22] Ansari, R., & Rouhi, H. (2012). Analytical treatment of the free vibration of single-walled carbon nanotubes based on the nonlocal Flugge shell theory. Journal of Engineering Materials and Technology, 134(1).

[23] Storch, J. A., & Elishakoff, I. (2013). Analytical solutions to the free vibration of a doublewalled carbon nanotube carrying a bacterium at its tip. Journal of Applied Physics, 114(17), 174309.

[24] Alibeigloo, A., & Shaban, M. (2013). Free vibration analysis of carbon nanotubes by using three-dimensional theory of elasticity. Acta Mechanica, 224(7), 1415–1427.

[25] Chawis, T., Somchai, C., & Li, T. (2013). Nonlocal theory for free vibration of single-walled carbon nanotubes, Advanced Materials Research, 747, 257–260.

[26] Smalley, R. E., Li, Y., Moore, V. C., Price, B. C., Colorado, Jr, R., Schmidt, H. K., Hauge, R. H., Barron, A. R., & Tour, J. M. (2006). Single Wall Carbon Nanotube Amplification: En Route to a Type-Specific Growth Mechanism. J Am Chem Soc, 128, 15824–15829.

[27] Sanchez-Valencia, J, R., Dienel, T., Gröning, O., Shorubalko, I., Mueller, A., Jansen, M., Amsharov, K., Ruffieux, P., & Fasel, R., (2014). Controlled synthesis of single-chiral carbon nanotubes. Nature, 512, 61–64.

[28] Gafour, Y., Zidour, M., Tounsi, A., Heireche, H., & Semmah, A. (2013). Sound wave propagation in zigzag double-walled carbon nanotubes embedded in an elastic medium using nonlocal elasticity theory. Physica E: Low-dimensional Systems and Nanostructures, 48, 118–123.

[29] Wu, C. P., & Lai, W. W. (2015). Free vibration of an embedded single-walled carbon nanotube with various boundary conditions using the RMVT-based nonlocal Timoshenko beam theory and DQ method. *Physica E: Low-dimensional Systems and Nanostructures, 68*, 8–21.

[30] Rouhi. H., Bazdid-Vahdati, M, & Ansari, R. (2015). Rayleigh-Ritz vibrational analysis of multi-walled carbon nanotubes based on the nonlocal Flügge shell theory. Journal of Composites, 10.1155/2015/750392.

[31] Azimzadeh, Z., & Fatahi-Vajari, A. (2019). Coupled axial-radial vibration of single-walled carbon nanotubes via doublet mechanics. Journal of Solid Mechanics, 11(2), 323–340.

[32] Khosravi, F., Hosseini, S. A., & Norouzi, H. (2020). Exponential and harmonic forced torsional vibration of single-walled carbon nanotube in an elastic medium. Proceedings of the Institution of Mechanical Engineers, Part C: Journal of Mechanical Engineering Science, 234(10), 1928–1942.

[33] Wang, Z., & Hu, W. (2021). Chaos of a single-walled carbon nanotube resulting from periodic parameter perturbation. International Journal of Bifurcation and Chaos, 31(9), 2150130.

[34] Selim, M. M. (2022). Torsional vibration of irregular single-walled carbon nanotube incorporating compressive initial stress effects. Journal of Mechanics, 37, 260–269.

[35] Ahmed, A. M., & Rifai, A. M. (2021). Euler-Bernoulli and Timoshenko Beam Theories Analytical and Numerical Comprehensive Revision. *European Journal of Engineering and Technology Research, 6*(7), 20–32.

[36] Reddy, J. N., & Wang, C. M. (2009). A bending, buckling and frequency relationships between the Euler-Bernoulli and Timoshenko nonlocal beam theories. *Asian Journal of Civil Engineering (Building and Housing), 10*(3), 265–281.

[37] Tserpes, K.I. & Papanikos, P. (2005). Finite element modeling of single-walled carbon nanotubes. *Comp. Part B: Eng.*, 36, 468–477.

# 7

# Concluding Remarks/Summary/Future Recommendation

## 7.1 Conclusion

In the last part of this book, the overall conclusions of single-walled CNTs for vibrational analysis have been presented and validated with existing open text. Since the percentage of error is negligible, the models have been concluded as valid. The vibrations of the carbon nanotube that have been investigated are based on shell theory of Flügge method, orthotropic shell model, Galerkin's methodlogy, Sander's shell theory, and Euler theory formulation. To generate the fundamental natural frequencies of SWCNTs, computer software MATLAB is engaged.

Free vibrations are developed using Flügge shell theory and Rayleigh–Ritz method for the vibration of single-walled carbon nanotubes. The frequencies of three sorts of SWCNTs are conducted with two boundary conditions. The resulting frequencies are gained for length-to-diameter ratios. The natural frequency becomes more prominent for lower length-to-diameter ratios and diminished for higher ratios. Different indices of armchair (4, 4), (11, 11), zigzag (6, 0), (13, 0), and chiral (7, 4), (10, 6) are investigated. The frequencies increase on decreasing the ratio of length-to-diameter. Also the frequencies increase when index increases as in the case of armchair indices (4, 4), (11, 11), for varying indices (6, 0), (13, 0) and indices of chiral tube (7, 4), (10, 6). This behavior of frequency remains the same with increasing the ratio of length-to-diameter throughout the computation in this chapter. It is observed that frequencies are not affected by changing the structure of single-walled carbon nanotubes as armchair, zigzag, and chiral. Also the boundary condition has a momentous effect on the vibration of single-walled carbon nanotubes. These boundary conditions are kept on the edge of tube. It is noted that the frequency curves of C–SS boundary condition are the lowest

135

outcomes compared to C–C boundary condition. This is due to physical constraints of boundary conditions.

Here the governing equations of orthotropic shell model are solved by novel wave propagation approach, to write the system equations in eigen form. In case of stiffness of the material, as the stiffness increases, the frequency becomes surge for both boundary conditions. If the material is stiff, then the frequency increases. However, the thickness of the structure remains the same. If the pattern of curves bends a little bit, then it means that material is less stiff. The simply supported frequencies are greater than that of clamped-free. This phenomenon of vibration is seen for armchair, zigzag, and chiral SWCNTs in case of stiffness separately.

Furthermore, a new method is based on Donnell shell theory to investigate the vibration of armchair, zigzag, and chiral single-walled carbon nanotubes. This model contains both the effect of boundary conditions and height-to-diameter ratios. The wave propagation approach is engaged for decretise the governing equations in eigen form. This eigen form is solved through MATLAB software to obtain the fundamental frequencies of SWCNTs. The effect of boundary conditions and height-to-diameter ratios for frequency behavior is discussed. The frequency pattern with two boundary conditions seems to be parallel for armchair, zigzag, and chiral. The frequencies increase on increasing ratio of height-to-diameter. With higher index, the frequencies will be higher. The C–C frequencies are higher than those corresponding C–F conditions. For shorter tube and shorter chiral index, the frequency displacement between the curves of C–C and C–F boundary condition is large.

In addition, a numerical approach is developed for the vibration of SWC-NTs based on Sander's shell theory. The Galerkin's technique is used to extract the frequencies of CNTs in the form of eigen value. Some examples are presented in tabular form to compare the results. The accuracy is found for excellent convergence behavior. A detailed parametric study is displayed for the influence of Poisson's ratio on armchair, zigzag and chiral tubes with simply supported and clamped simply supported edge condition. It is seen that frequencies increase on increasing the Poisson's ratio. When inclusion of Poisson's effect is considered during vibration, then effective stiffness increases which increase the natural frequencies. The estimated frequency values of C–SS are high compared to SS–SS. The frequency value increases with the increase of indices of single-walled carbon nanotubes. The frequency pattern with all boundary conditions seems to be parallel for overall values of Poisson's ratio.

Moreover, the Euler beam theory is utilized to obtain the small size influence on the variation of density response for armchair, zigzag, and chiral single-walled carbon nanotubes. The frequencies of SWCNTs are instigated for aspect ratios and half-axial wave mode. It is noted that the frequencies of C–C are higher than those of C–F. This modified model has less complication and has been compared with the earlier methods. The computational results indicated that there is an inverse relation of aspect ratios and axial wave mode frequencies. The frequency curves of clamped-free are lower throughout the computation than other boundary conditions as applied at the end of nanotube. The obtained results show that by increasing aspect ratio of carbon nanotubes, frequency decreases at all boundary conditions. In present measurement, it is indicated that with higher aspect ratio, the BCs have a momentous influence on vibration of CNT. It can be concluded that frequencies would increase by the increase of half-axial wave modes.

In summary, a new model is developed for the vibration of armchair, zigzag, and chiral single-walled carbon nanotubes with effect of length-to-diameter ratios, stiffness, height-to-diameter ratios, and Poisson's ratio. It is necessary to see the behavior of armchair, zigzag, density, and chiral tubes, because in vibration, the deformations occur. So this type of deformation is key type for the strength of the material and it can avoid material loss. Hence, this study is a very powerful tool for organizing material in tiny instruments, sensor, and actuators. The variation in stiffness is of significant importance for designing and performance of the structures. In the materials, Poisson's ratio directly measures the deformation and material shows a large elastic deformation due to high Poisson's ratio. When the Poisson's ratio increases, the moduli are expected to become so high and as a result sufficiently high frequencies are observed. In stretching and compression, Poisson's ratio is a useful measure of how much a material is deformed. It is important for mechanical engineering as it allows the materials to be chosen that suit the desired function. The validation of the present numerical approach with large amount of exact and experimental calculations such as with beam element, Timoshenko beam model, molecular dynamic (MD) simulations, DTM and Bubnov–Galerkin method, Resonant Raman Scattering (RRS), and Raman Spectroscopy depicts the great assurance of the present established model for practical engineering. The author expects this frequency analysis for high frequencies in fascinating electromagnetic devices. The theoretical results obtained from the paper can be used for the sensitivity analysis of the nano- and bio-sensors.

## 7.2 Future Recommendation

For future work, the vibration analysis of the nanostructures is a new and very hot topic, and accordingly, there are many unsolved and vague problems for the future researchers in this area. These shell theories can be utilized for analyzing the vibrations in carbon nanotube by considering different parameters, namely, Winkler and Pasternak foundations, geometrical imperfections, fluid conveying, and thermal and electromagnetic effects. The theoretical results obtained from the book can be used for the sensitivity analysis of the nano- and bio-sensors. The proposed models of the present work can be used for the vibration analysis with experimental set. The vibrations of the piezoelectric-based SWCNT resonators can be investigated using the developed theory of the book for the curved structure and piezoelectricity based on FGM carbon nanotubes can be also investigated as a new topic for the future researchers. The electrostatic and van der Waals can highly influence the vibrations of the CNTs. Accordingly, it would be interesting to consider the above forces for the developed models of book or for the MWCNTs in future research.

# Appendices

## Appendix 2.1

Where

$$A_{11} = \frac{Eh}{1 - v^2}$$

$$A_{12} = \frac{vEh}{1 - v^2}$$

$$A_{22} = \frac{Eh}{1 - v^2}$$

$$A_{66} = \frac{Eh}{2(1 + v)}, A_{16} = A_{26} = 0$$

$$D_{11} = \frac{Eh^3}{12(1 + v)}$$

$$D_{12} = \frac{vEh^3}{12(1 + v)}$$

$$D_{22} = \frac{Eh^3}{12(1 - v^2)}$$

$$D_{66} = \frac{Eh^3}{24(1 + v)}, D_{16} = D_{26} = 0$$

## Appendix 2.2

$$C_{11} = A_{11} \int_0^L \left(\frac{d^2\varphi}{dx^2}\right)^2 dx + \frac{n^2 A_{66}}{R^2} \int_0^L \left(\frac{d\varphi}{dx}\right)^2 dx$$
$$+ \frac{4n^2 B_{66}}{R^2} \int_0^L \left(\frac{d\varphi}{dx}\right)^2 dx + \frac{4n^2 D_{66}}{R^4} \int_0^L \left(\frac{d\varphi}{dx}\right)^2 dx$$

$$C_{12} = \left( -\frac{2nA_{12}}{R} \int_0^L \left( \frac{d^2\varphi}{dx^2} \right) \varphi(x)dx + \frac{2nA_{66}}{R} \int_0^L \left( \frac{d\varphi}{dx} \right)^2 dx \right.$$

$$\left. -\frac{8nD_{66}}{R^3} \int_0^L \left( \frac{d\varphi}{dx} \right)^2 dx \right)^2$$

$$C_{13} = \frac{2A_{12}}{R} \int_0^L \left( \frac{d^2\varphi}{dx^2} \right) \varphi(x)dx - 2B_{11} \int_0^L \left( \frac{d^2\varphi}{dx^2} \right)^2 dx$$

$$-\frac{2B_{12}}{R^2} \left( -n^2 \int_0^L \left( \frac{d^2\varphi}{dx^2} \right) \varphi(x)dx + \int_0^L \left( \frac{d^2\varphi}{dx^2} \right) \varphi(x)dx \right)$$

$$+\frac{4n^2 B_{66}}{R^2} \int_0^L \left( \frac{d\varphi}{dx} \right)^2 dx + \frac{16n^2 D_{66}}{R^3} \int_0^L \left( \frac{d\varphi}{dx} \right)^2 dx$$

$$C_{21} = \left( -\frac{2nA_{12}}{R} \int_0^L \left( \frac{d^2\phi}{dx^2} \right) \phi(x)dx + \frac{2nA_{66}}{R} \int_0^L \left( \frac{d\phi}{dx} \right)^2 dx \right.$$

$$\left. -\frac{8nD_{66}}{R^3} \int_0^L \left( \frac{d\phi}{dx} \right)^2 dx \right)$$

$$C_{22} = \left( \frac{n^2 A_{22}}{R^2} \int_0^L \varphi^2(x)dx + \frac{A_{66}}{R} \int_0^L \left( \frac{d\varphi}{dx} \right)^2 dx \right.$$

$$\left. -\frac{4B_{66}}{R} \int_0^L \left( \frac{d\varphi}{dx} \right)^2 dx + \frac{4D_{66}}{R^2} \int_0^L \left( \frac{d\varphi}{dx} \right)^2 dx \right)$$

$$C_{23} = -\frac{2nA_{22}}{R^2} \int_0^L \varphi^2(x)dx + \frac{2nB_{12}}{R} \int_0^L \frac{d^2\varphi}{dx^2} \varphi(x)dx$$

$$-\frac{2B_{22}}{R^3} \left( n^3 \int_0^L \varphi^2(x) - n \int_0^L \varphi^2(x) \right)$$

$$+\frac{8nB_{66}}{R} \int_0^L \left( \frac{d\varphi}{dx} \right)^2 dx + \frac{16nD_{66}}{R^2} \int_0^L \left( \frac{d\varphi}{dx} \right)^2 dx$$

$$C_{31} = \frac{2A_{12}}{R} \int_0^L \left( \frac{d^2\varphi}{dx^2} \right) \varphi(x)dx - 2B_{11} \int_0^L \left( \frac{d^2\varphi}{dx^2} \right)^2 dx$$

$$-\frac{2B_{12}}{R^2} \left( -n^2 \int_0^L \left( \frac{d^2\varphi}{dx^2} \right) \varphi(x)dx + \int_0^L \left( \frac{d^2\varphi}{dx^2} \right) \varphi(x)dx \right)$$

$$+\frac{4n^2 B_{66}}{R^2} \int_0^L \left( \frac{d\varphi}{dx} \right)^2 dx + \frac{16n^2 D_{66}}{R^3} \int_0^L \left( \frac{d\varphi}{dx} \right)^2 dx$$

$$C_{32} = -\frac{2n\,A_{22}}{R^2} \int_0^L \varphi^2(x)dx + \frac{2nB_{12}}{R} \int_0^L \frac{d^2\psi}{dx^2}\varphi(x)dx$$

$$-\frac{2B_{22}}{R^3}\left(n^3\int_0^L \varphi^2(x) - n\int_0^L \varphi^2(x)\right)$$

$$+\frac{8nB_{66}}{R}\int_0^L \left(\frac{d\varphi}{dx}\right)^2 dx + \frac{16nD_{66}}{R^2}\int_0^x \left(\frac{d\varphi}{dx}\right)^2 dx$$

$$C_{33} = \frac{A_{22}}{R^2}\int_0^L \varphi^2(x)dx - \frac{2B_{12}}{R}\int_0^L \frac{d^2\varphi}{dx^2}\varphi(x)dx$$

$$-\frac{2B_{22}}{R^3}\left(-n^2\int_0^L \varphi^2(x)dx + \int_0^L \varphi^2(x)dx\right)+$$

$$D_{11}\int_0^L \left(\frac{d^2\varphi}{dx^2}\right)^2 dx + \frac{D_{22}}{R^4}\left(n^4\int_0^L \varphi^2(x)dx + \int_0^L \varphi^2(x)dx\right)$$

$$-2n^2\int_0^L \varphi^2(x)dx\right)$$

$$+\frac{2D_{12}}{R^2}\left(-n^2\int_0^L \frac{d^2\varphi}{dx^2}\varphi(x)dx + \int_0^L \frac{d^2\varphi}{dx^2}\varphi(x)dx\right)$$

$$+\frac{16n^2D_{66}}{R^2}\int_0^2 \left(\frac{d^2\varphi}{dx^2}\right)^2 dx$$

## Appendix 3.1

Where

$$A_{11} = -k_m^2 - n^2k_2\left(1+c^2\right)$$
$$A_{12} = n\left(\mu_1 + k_2\right) - ik_m$$
$$A_{13} = -\mu_1 ik_m + n^2c^2k_2 ik_m - ik_m^3$$
$$A_{21} = n\left(\mu_1 + k_2\right)k_m$$
$$A_{22} = -k_2\left(1+3c^2\right)k_m^2 - n^2k_1$$
$$A_{23} = -nk_1 - nc^2\left(\mu_1 + 3k_2\right)k_m^2$$
$$A_{31} = -\mu_1 ik_m - c^2\left(ik_m^3 - n^2k_2 ik_m\right)$$
$$A_{32} = nk_1 + nc^2\mu_1 k_2 k_m^2$$
$$A_{33} = \left(1+\frac{1}{c^2}\right)k_1 + k_m^3 + n^4k_1 - 2n^2k_1 + \left(2\mu_1 + 4k_2\right)n^2k_m^2$$

## Appendix 4.1

Where

$$\Delta_{11} = l^2{}_c + \frac{1-v}{2R^2}n^2$$

$$\Delta_{12} = il_c\frac{1+v}{2R}n$$

$$\Delta_{13} = il_c\frac{v}{R}$$

$$\Delta_{21} = -n\frac{1+v}{2R}il_c$$

$$\Delta_{22} = \frac{1-v}{2}l_c^2 + \frac{n^2}{R^2}$$

$$\Delta_{23} = \frac{n}{R^2}$$

$$\Delta_{31} = -\frac{v}{R}il_c$$

$$\Delta_{32} = \frac{n}{R^2}$$

$$\Delta_{33} = \frac{1}{R^2} + \frac{(1-v^2)}{Eh}D\left(l_c^4 + 2\frac{1}{R^2}l_c^2n^2 + \frac{n^4}{R^4}\right)$$

# Index

**V**

Vibration 31, 35, 43, 56, 78, 89, 105, 117, 126, 125

**W**

Wave propagation approach 7, 33, 54, 120

**Z**

Zigzag 7, 30, 33, 58, 60, 77, 102, 121, 127, 136

# About the Author

**Dr. Muzamal Hussain** received his Ph.D. degree in Computational Mathematics from GC University Faisalabad, Pakistan in 2019. His main research areas include computational mathematics, vibration of a rotating functionally graded cylindrical shell with various volume fraction laws, and vibration analysis of single-walled carbon nanotubes based on different theories using the wave propagation approach. He has published more than 150 research papers and many book chapters in international impact factor journals and citations of his articles more than 2000.

Due to these research excellence, author name is included as a young scientist in many research societies/index.

Printed in the United States
by Baker & Taylor Publisher Services